Simple Guide on Management and Control of Wastes

Simple Guide on Management and Control of Wastes

Study carried out by the Royal Society of Chemistry, London, for the European Commission, under Contract B4-3040-93-686

This study was carried out by a specialist committee of the Royal Society of Chemistry for the European Commission, Directorate-General XI, Directorate A: Environment, Nuclear Safety and Civil Protection. The members of the committee were:

Mr G J Dickes
Dr J H Duffus
Mr T G R Farthing
Mr R W Hazell (Secretary)
Mr S G Luxon (Chairman)
Mr D M Sanderson
Mr P A Stanbridge
Dr D Taylor

The Royal Society of Chemistry would like to acknowledge the assistance of the following organisations in giving information.

The American Chemical Society
The Department of the Environment [UK]
The European Foundation for the Improvement of Living and
    Working Conditions
The International Solid Wastes and Public Cleansing Association
The National Federation of Women's Institutes [England and Wales]
The United Nations Environment Programme
The World Health Organisation

A catalogue record for this book is available from the Brithish Library.

ISBN 0-85404-990-8

Printed and bound by Redwood Books Ltd., Trowbridge, Wiltshire

Contents

Glossary of Acronyms

ACS	American Chemical Society
BATNEEC	Best available technology not entailing excessive cost
BOD	Biochemical oxygen demand
CEC	Commission of the European Communities
CEFIC	European Council of Chemical Manufacturers' Federations
COD	Chemical oxygen demand
DDT	Dichlorodiphenyltrichloroethane
EC	European Community
ECETOC	European Centre for Ecotoxicology of Chemicals
EEC	European Economic Community
EFTA	European Free Trade Association
EIA	Environmental impact assessment
EU	European Union
IPC	Integrated pollution control
ISWA	International Solid Waste and Public Cleansing Association
LCA	Life cycle analysis
OECD	Organisation for Economic Co-operation and Development
PCBs	Polychlorinated biphenyls
PCDDs	Polychlorinated dibenzo-p-dioxins
PCDFs	Polychlorinated dibenzofurans
PCTs	Polychlorinated terphenyls
TOC	Total organic carbon
UNEP	United Nations Environment Programme
WHO	World Health Organisation

Executive Summary

This publication has been compiled to identify the major moral, legal and administrative duties associated with the minimisation, control, and ultimate disposal of liquid, semi-solid, and solid waste, in a manner which could be said to represent the current ideas of "Good Environmental Practice".

No single publication of this nature could hope to address the means by which all waste streams to the three media – air, water, and land – need to be controlled. There are, for example, certain waste materials which require specific treatment and disposal methods which are outside the scope of this guide – gaseous emissions, radioactive waste, mining waste, agricultural waste (including carcasses and silage liquor), and sewage.

The major objective has been to inform, in a technical manner, rather than to educate, in the scientific sense. Consequently, processes are identified but not detailed. This guide will:

- provide a reference work for those who need to acquaint themselves with the principles of waste management
- satisfy the requirements of people who wish to have a deeper understanding of the current international waste controls in commerce and industry

The work is divided into three chapters; Introduction, Technical Aspects, and Legal/Administrative/Economic Aspects. Each chapter is sub-divided with headings to assist the reader in locating specific information.

The Introduction commences with a survey of the means by which waste may be adequately defined. It identifies the threats to public health and the quality of the environment posed by mismanagement of waste, and concludes with an outline strategy in relation to the public perception of waste management. Throughout, the recommendations of the EU, which represent the considered views of twelve major States, have been used as examples.

It is now clear that the amount of waste from the activities of day-to-day living, commerce and industry must be severely curtailed.

Waste material can result from the inefficiency of an activity. It may, in some cases, be an unavoidable burden to be borne by a manufacturer if he is to maintain his product line. However, with the increasing cost of waste disposal for a variety of technical, environmental and political reasons, the attitude that all waste is regrettable but inevitable can no longer be afforded.

The strategy for the future must be to minimise waste production by design, to recycle where possible, and ensure that disposal is efficient and is carried out in an environmentally acceptable manner.

Under the title 'Technical Aspects' the theme is developed. It is shown that effective waste minimisation in an organisation or State can only be achieved by the commitment of all of its members. That means recognition by all concerned that waste reduction is a worthwhile activity, and that recycled goods are just as good as those manufactured from virgin raw materials. In certain cases, this will entail the provision of incentives, and commendation for excellent performance.

The benefits and opportunities afforded by a well-managed recycling programme are discussed; specific examples are provided of the means of reducing hazardous waste by substitution. The importance is stressed of the necessity to characterise waste accurately in order to ensure its safe handling, storage and transportation.

The industrial or commercial waste producer is clearly in possession of the essential details from which general characteristics may be obtained. However, for some industrial wastes, more specific chemical information on the nature and concentration of hazardous constituents will be required in order that the ecologically correct treatment and/or disposal method will be selected. The general methods by which this may be accomplished are outlined.

Details are provided of the Basel Convention. This was formulated to prevent the future indiscriminate export of hazardous or intractable waste to unsuspecting countries which are ill-prepared to handle it.

The newcomer to waste treatment will welcome the list of processes used to reduce the bulk of a waste stream and/or render it more ecologically acceptable for ultimate disposal. The processes outlined cover physical, chemical and biological methods which are currently used individually, or in combination, according to the nature and character of the waste. These processes constitute, in general, a large proportion of the costs of waste management, due to the relatively high cost of installation and operation of plant and services to support it. The chapter closes with a similar overview of the major waste disposal methods.

Incineration is virtually the only option for certain types of waste, particularly synthetic materials originally formulated for their chemical and biological stability. The various types of incineration plant in current use are described with reference to the regulatory controls for ecological protection and safe operation of these devices.

Landfill is the final resting place for the major proportion of waste substances, whether directly, or following pre-treatment. In order to protect the surrounding area from contamination, the problems of maintaining a new ecologically acceptable landfill have to be addressed by investigating the geology and hydrogeology of the site before the first deposit of waste is made.

In the past, the deposition of waste to landfill was not entered into with anything like this degree of preparation. Control and treatment of gaseous emissions (mainly methane, due to microbiological activity), and liquid leachate (due to the combined effects of rainfall and microbiological activity), are concisely reported.

In the final chapter, the national administrative framework for waste management is addressed. It is clear that in order to protect future generations of the human population together with flora and fauna, the disposal of waste must be carried out according to a carefully managed plan. The ways in which such national plans should be formulated is described. Localities within a State must also introduce systems which will be compatible with the parent plans. Such plans and systems will place economic and moral responsibilities upon industry, commerce and the individual to review continually for possible reduction of the quantity of waste produced.

The means by which waste is disposed of will be subject to ever more stringent controls in the form of permits or licences, the contractual conditions of which will be carefully audited for strict compliance. The recently formulated system of life-cycle analysis is summarised. This method, if used prudently, will enable the manufacturer or producer to meet more easily such targets for waste reduction as may be set locally or nationally to achieve the goals set in the national/local plans.

Life-cycle analysis is an overall audit of the ecological compatibility of a product, taking into account raw material production, all parts of the manufacturing process, the lifetime operation of the product, and its ultimate disposal, including the pollution associated with energy usage at all stages.

The text is supported by data in diagrammatical form, together with tables of the EU waste identification criteria.

Sequence of waste treatment processes

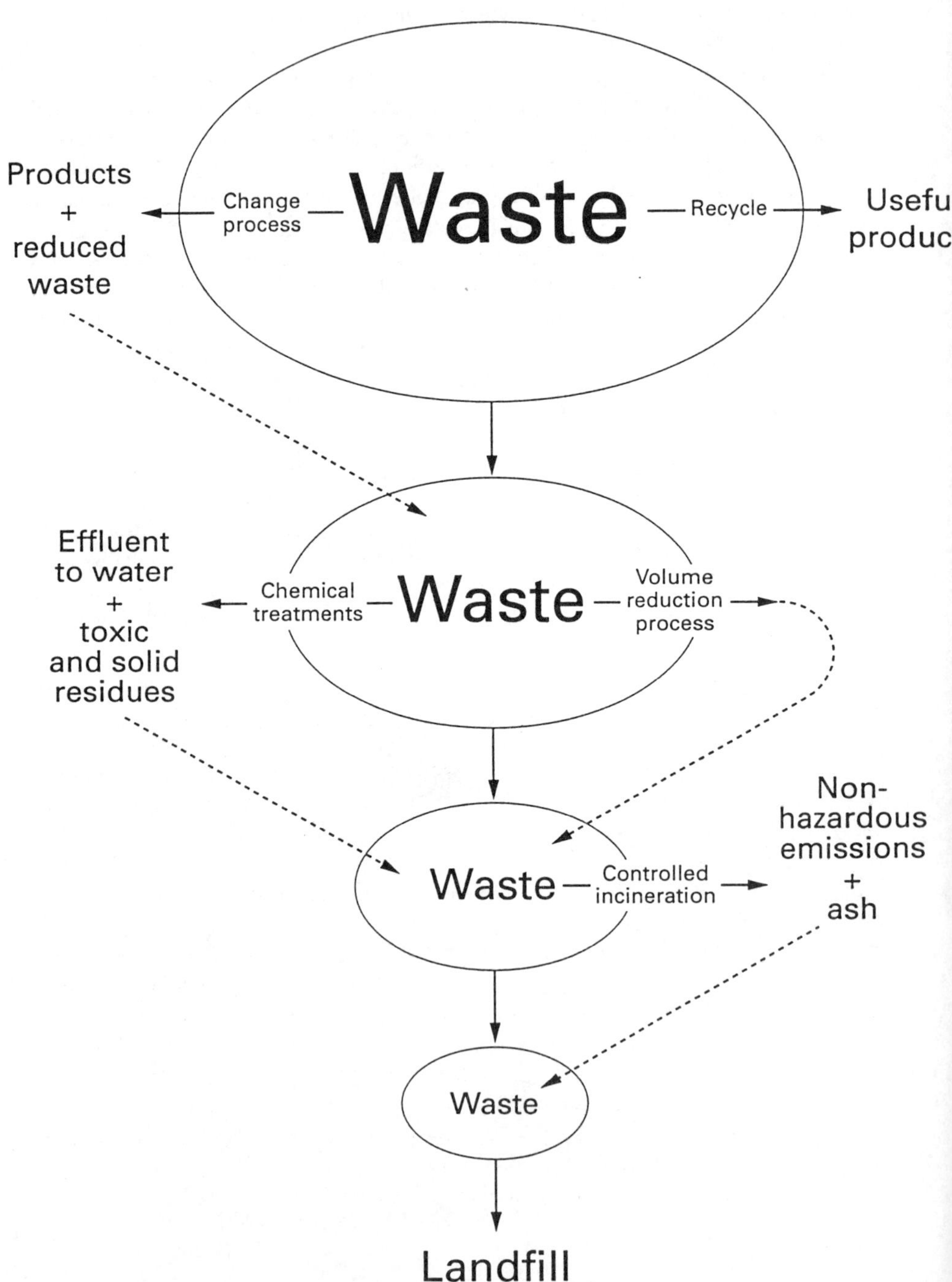

Chapter I

Introduction

1 Scope

This is a simple guide and is not intended as a treatise on waste management. Although there is emphasis on EC philosophy of waste management, and therefore on relevant EC legislation, the guide is meant for all and for developing countries in particular. This Community philosophy embodies the distilled wisdom of twelve countries.

EC waste legislation has developed since the 1970s and this is illustrated in FIGURE 1. This schematic diagram is expanded in TABLE 1.

Integrated pollution control (IPC) is practised in the Community and means that the management of waste cannot be considered in isolation. The risks of air and groundwater pollution from waste disposal activities have to be taken into consideration as well as any risks from the waste itself. There are wastes, not covered in this guide, the polluting effects of which have to be taken into account in IPC. These excluded waste materials, in line with EC Directive 91/156/EEC, are: gaseous effluents, radioactive wastes, mining wastes, animal carcasses, faecal matter and other natural non-dangerous substances used in farming, waste waters (including sewage). For reasons of continuity, gaseous effluents and sewage are sometimes mentioned in the text.

In order to prevent or minimise waste, the cleanest and most economic production technologies are advocated. The underlying principle applying to all potentially polluting activities is termed "best available technology not entailing excessive cost" (BATNEEC).

BATNEEC appears to place the sole onus of responsibility for attaining a cleaner environment on the shoulders of industry. However, everyone has a responsibility, from Government to the individual. The environmental movement has used public concern to press for legislation to regulate potentially polluting activities, such as waste disposal. Environmentalists have had greatest impact after reasoned, scientifically-based assessments of risk with the acceptance that some elements of waste have to reside somewhere, finally.

Legal enforcement and environmental lobbying cannot achieve satisfactory results on their own. There needs to be co-operation between politicians, industry and public to establish safer ways of disposing of waste, once the alternatives of waste prevention, re-use and recycling have been exhausted.

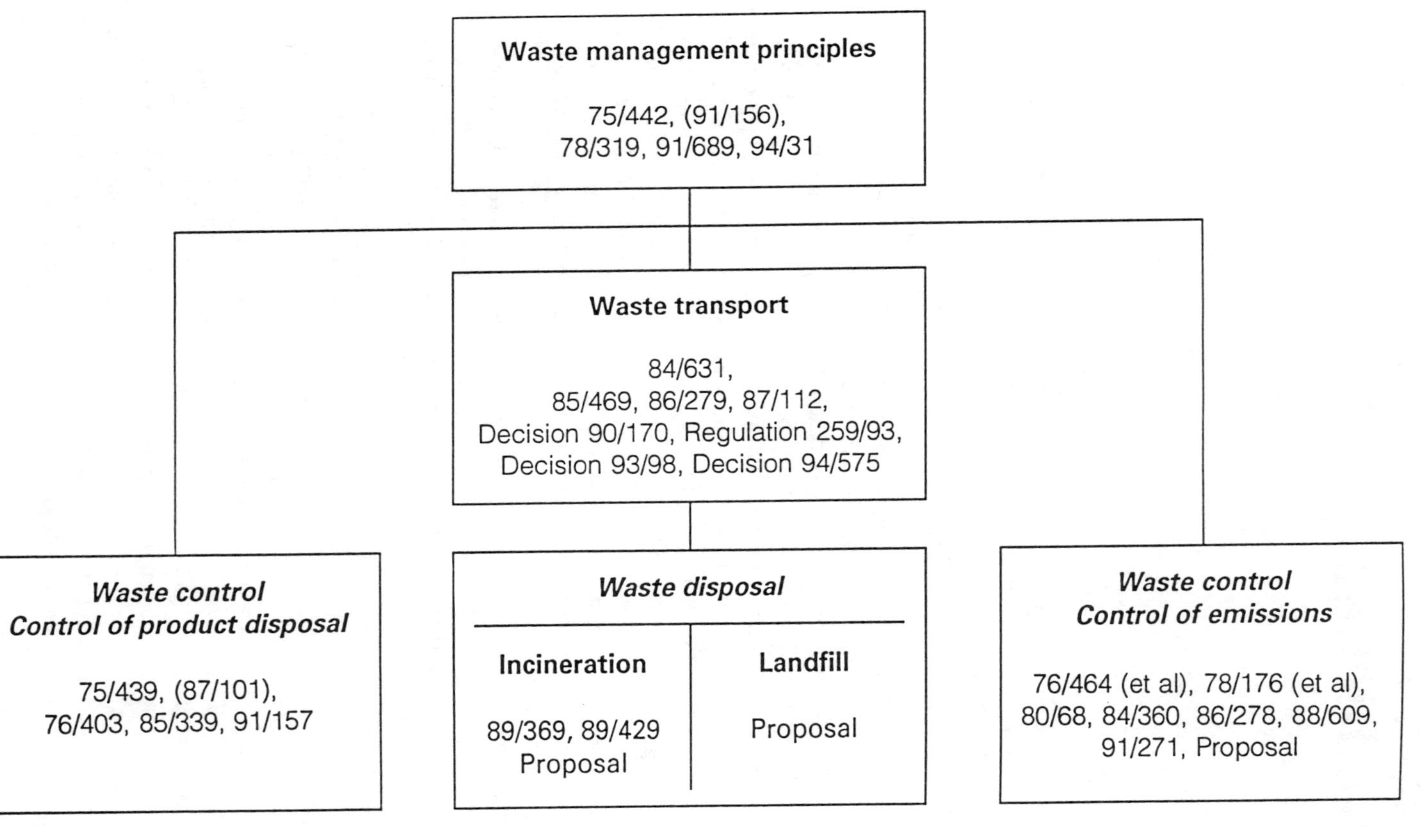

Figure 1 *European Union directives related to waste management*

Table 1 *List of EC Directives related to waste management*

Initial Directive	Subsequent amendments	Title
Waste Management		
75/442	91/156	on waste
78/319		on toxic & dangerous waste
91/689		on hazardous waste
Waste Transport		
84/631	85/469, 86/279 87/112 Decision 90/170 Regulation 259/93 Decision 94/721	on supervision & control within the EC of the transfrontier shipment of hazardous waste
Waste Treatment & Disposal		
89/429		on the reduction of air pollution from existing municipal waste incinerators
89/369		on the reduction of air pollution from new municipal waste incinerators
94/67		on hazardous waste incineration
COM(93)275		proposal on the landfill of waste
Waste Products		
75/439	87/101	on the disposal of waste oils
76/403		on the disposal of PCBs and PCTs
85/339		on containers of liquids for human consumption
91/157		on batteries and accumulators containing certain dangerous substances
Waste Emissions		
76/464	82/176, 83/513 84/156, 84/491 86/280, 88/347 90/415	on the discharge of dangerous substances
78/176	80/779, 84/360 89/428, 92/112	on waste from the titanium dioxide industry
80/68		on the protection of groundwater against pollution by certain dangerous substances
86/278		on the protection of the environment when sewage sludge is used in agriculture
91/271		concerning urban waste water treatment
95/ENV/154		proposal on integrated pollution prevention & control

2 Definition of waste

There are many different definitions of waste and of types of waste throughout the world. The interpretation of what is meant by waste can also change with circumstances, e.g. what is waste to one can be a useful raw material to another; and waste food for human consumption can be used as animal fodder.

2.1 EC definition of waste and the European Waste Catalogue

The first EC Directive specifically dealing with waste is 75/442/EEC, which is a framework Directive which has been filled out in parts since its inception. Waste is defined here as any substance or object which the holder disposes of or is required to dispose of pursuant to the provisions of national law in force.

In an acknowledgement that some Member States had already categorised waste, the Directive allows that without prejudice to this Directive, Member States may adopt specific rules for particular categories of waste. There are other definitions of waste throughout the world but these have not been considered in this guide.

In the EC Directive, 91/156/EEC, which is largely an amended version of the framework Directive, waste is re-defined to mean any substance in the categories set out in Annex I of the Directive which the holder discards or intends or is required to discard. It could be assumed that if a material is not covered by any of these categories, it is not a waste as defined by the Directive.

Provision was made in the Directive for the Commission to draw up a list of wastes divided into the sixteen categories of Annex I, and this list, commonly known as the European Waste Catalogue, was issued on 20 December 1993. The Catalogue applies to all wastes, irrespective of whether they are destined for disposal or for recovery; it is a harmonised, non-exhaustive catalogue of 645 wastes, which will be periodically reviewed and amended as necessary. The original categories of Annex I to 91/156/EEC have been re-issued in a slightly different form and number twenty in the European Waste Catalogue, and these are shown in TABLE 2.

Table 2 *Index of the European waste catalogue*

(Commission decision of 20th December 1993, establishing a list of wastes puruant to Article 1(a) of Council Directive 75/442/EEC)

---

1 Waste resulting from exploration, mining, dressing and further treatment of minerals and quarrying
2 Waste from agricultural, horticultural, hunting, fishing and aquaculture primary production, food preparation and processing
3 Wastes from wood processing and the production of paper, cardboard, pulp, panels and furniture
4 Wastes from the leather and textile industries
5 Wastes from petroleum refining, natural gas purification and pyrolytic treatment of coal
6 Wastes from inorganic chemical processes

Table 2 *continued*

---

7 Wastes from organic chemical processes
8 Wastes from the manufacture, formulation, supply and use of coatings (paints, varnishes and vitreous enamels), adhesives, sealants and printing inks
9 Wastes from the photographic industry
10 Inorganic wastes from thermal processes
11 Inorganic waste with metals from metal treatment and the coating of metals; non-ferrous hydrometallurgy
12 Wastes from shaping and surface treatment of metals and plastics
13 Oil wastes (except edible oils and categories 5 and 12)
14 Wastes from organic substances employed as solvents (except categories 7 and 8)
15 Packaging; absorbents, wiping cloths, filter materials and protective clothing not otherwise specified
16 Waste not otherwise specified in the catalogue
17 Construction and demolition waste (including road construction)
18 Wastes from human or animal health care and/or related research (including kitchen and restaurant wastes which do not arise from immediate health care)
19 Wastes from waste treatment facilities, off-site waste water treatment plants and the water industry
20 Municipal wastes and similar commercial, industrial and institutional wastes, including separately collected fractions

---

The Catalogue is a reference system of nomenclature and criteria in order to compare the efficiency of waste management activities. The Catalogue should constitute the basic reference for the European Community Programme on waste statistics launched pursuant to the European Council Resolution, 90/C 122/02 on waste management policy.

The inclusion of a material in the Catalogue does not mean that the material is a waste in all circumstances. The entry is only relevant when the definition of waste has been satisfied.

2.2 Toxic, dangerous and hazardous wastes and the EC list of hazardous wastes

The EC Directive on toxic and dangerous waste, 78/319/EEC, defines toxic and dangerous waste as any waste containing or contaminated by the substances or materials listed in the Annex to the Directive, of such a nature, in such quantities or in such concentrations as to constitute a risk to health or the environment. The Annex contains 27 toxic or dangerous substances and materials selected as requiring priority consideration.

This Directive will be superseded by the EC Directive 91/689/EEC, as modified by Directive 94/31, on hazardous waste, which defines hazardous wastes as those featuring in a list to be drawn up in accordance with the procedure laid down in the framework Directive and based on Annexes I and II to the present Directive, provided they possess one or more of the hazardous properties listed in Annex III.

Annex I, which is divided into A and B, lists the categories of generic types of hazardous waste according to their nature or the activity which generated them. Annex IA lists 18 wastes which are hazardous if displaying any of the properties listed in Annex III and Annex 1B lists 22 wastes which are hazardous if containing any of the constituents listed in Annex II, provided they have any of the properties listed in Annex III. Annexes 1A and 1B are shown in TABLES 3A and 3B respectively.

Table 3A *Annex IA of EC Directive 91/689/EEC. Categories or generic types of hazardous wastes listed according to their nature or the activity which generated them*

ANNEX 1A   *Wastes displaying any of the properties listed in Annex III and which consist of:*

anatomical substances, hospital and other clinical wastes
pharmaceuticals, medicines and veterinary compounds
wood preservatives
biocides and phytopharmaceutical subtances
residue from substances employed as solvents
halogenated organic substances not employed as solvents and not inert polymerised
    materials
tempering salts containing cyanides
mineral oils and oily substances
oil/water, hydrocarbon/water mixtures, emulsions
substances containing PCBs and/or PCTs
tarry materials arising from refining, distillation and any pyrolytic treatment
inks, dyes, pigments, paints, lacquers, varnishes
resins, latex, plasticisers, glues/adhesives
chemical substances arising from research and development or teaching activities which are
    not identified and/or are new and whose effects on man and/or the environment are not
    known (e.g. laboratory residues, etc.)
pyrotechnics and other explosive materials
photographic chemicals and processing materials
any material contaminated with any congener of PCDF
any material contaminated with any congener of PCDD

Table 3B *Annex IB of EC Directive 91/689/EEC. Categories of generic types of hazardous wastes listed according to their nature or the activity which generated them*

ANNEX 1B *Wastes which contain any of the constituents listed in Annex II and having any of the properties listed in Annex III and consisting of:*

animal or vegetable scraps, fats, waxes
non-halogenated organic substances not employed as solvents
inorganic substances with metals or metal compounds
ashes and/or cinders
soil, sand, clay, including dredging spoils

Table 3B *continued*

non-cyanidic tempering salts
metallic dust, powder
spent catalyst materials
liquids or sludges containing metal or metal compounds
residues from pollution control operations
scrubber sludges
sludges from water purification plants
decarbonisation residue
ion-exchange column residue
sewage sludges, untreated or unsuitable for use in agriculture
residue from cleaning of tanks and/or equipment
contaminated equipment
contaminated containers whose contents included one or more of the constituents listed in
  Annex II
batteries and other electrical cells
vegetable oils
materials resulting from selective waste collections from households and which exhibit any
  of the characteristics listed in Annex III
any other wastes which contain any of the constituents listed in Annex II and any of the
  properties listed in Annex III

Annex II is a list of 53 constituents that, when present in the wastes listed in Annex 1B, render them hazardous, provided they have any of the properties listed in Annex III. These constituents are comprised largely of chemicals which have an established record of being hazardous, and these are shown in TABLE 4.

Table 4 *Annex II of EC Directive 91/689/EEC. Constituents of the wastes in Annex 1B which render them hazardous when they have the properties described in Annex III*

Beryllium and its compounds	Peroxides
Vanadium compounds	Chlorates
Chromium VI compounds	Perchlorates
Cobalt compounds	Azides
Nickel compounds	PCBs and/or PCTs
Copper compounds	Pharmaceutical or veterinary compounds
Zinc compounds	Biocides and phytopharmaceutical substances
Arsenic and its compounds	Infectious substances
Selenium and its compounds	Creosotes
Silver compounds	Isocyanates; thiocyanates
Cadmium and its compounds	Organic cyanides
Tin compounds	Phenols; phenol compounds
Antimony and its compounds	Halogenated solvents
Tellurium and its compounds	Organic solvents (non-halogenated)
Barium compounds (not sulfate)	Organohalogen compounds, excluding inert
Mercury and its compounds	polymerised materials and other
Metal carbonyls	substancesinert referred to in this Annex

Table 4   *continued*

Thallium and its compounds	Aromatic compounds, polycyclic and
Lead and its compounds	heterocyclic organic compounds
Inorganic sulfides	Aliphatic amines
Inorganic fluorine compounds (not	Aromatic amines
calcium fluoride)	Ethers
Inorganic cyanides	Substances of an explosive character
Lithium, sodium, potassium, calcium,	Sulfur organic compounds
magnesium	Any congener of PCDF
Acidic solutions or acids in solid form	Any congener of PCDD
Basic solutions or bases in solid form	Hydrocarbons and their oxygen, hydrogen
Asbestos (dust and fibres)	and/or sulfur compounds
Phosphorus and its compounds (not	
mineral phosphates)	

The properties of waste which render it hazardous (Annex III) are shown in TABLE 5.

Table 5   *Summary of **Annex III** of EC Directive 91/689/EEC. Properties of wastes which render them hazardous*

Explosive
Oxidising
{ Highly flammable
  Flammable
Irritant*
Harmful*
Toxic*
Carcinogenic*
Substances and preparations capable of yielding very toxic or toxic gases in contact with
   water, air, acid
Substances and preparations capable, after disposal, of yielding another substance, e.g.
   leachate, which possesses harmful characteristics
Corrosive*
Infectious
Teratogenic*
Mutagenic*
Ecotoxic

* Criteria laid down in EC Directive 67/548/EEC

A Member State may submit details to the Commission of any other waste for inclusion in Annex I if it is considered that the waste has any of the hazardous properties listed in Annex III.

2.3 The shipments of waste

The framework EC Directive 75/442/EEC lays down that an integrated and adequate network of waste disposal installations be established by Member States through appropriate measures, where necessary or advisable in co-operation with other Member States, and must enable the Community as a whole to become self-sufficient in waste disposal. This co-operation includes Member States being able to take measures to prevent movements of waste which are not in accordance with their waste management plans. Otherwise, with strict practical and documentary control, shipments of waste are allowed within the Community.

The primary objective of EC Council Regulation 259/93/EEC is to improve the system of supervision and control of all waste transfers within the European Community. Being a Regulation, its provisions have to be immediately enforced by Member States, tightening up the provisions of EC Directive 84/631/EEC on the supervision and control within the European Community of the transfrontier shipments of hazardous waste.

The Regulation lists three different types of waste: green, which is non-hazardous; red, which is very hazardous; amber, hazardous and includes waste which is under review, possibly to move it into one of the other two categories.

The main provisions of the Regulation are covered in Chapter II 3, Transboundary movement of waste.

3 The need to control waste

3.1 Risk to humans

If waste management is carried out properly, there is very little risk to human health. All waste, particularly hazardous waste, needs to be under control from initial collection to final disposal. In the framework EC Directive 75/442/EEC on waste, it is stated that the essential objective of all provisions relating to waste disposal must be the protection of human health and the environment against harmful effects caused by the collection, transport, treatment, storage and tipping of waste. Article 4 of the Directive states that Member States shall take the necessary measures to ensure that waste is disposed of without endangering human health and without harming the environment.

The control of municipal waste is the consumers' responsibility until it is collected, whereupon it becomes the responsibility of the collecting authority. In normal circumstances, the public should not come into contact with municipal waste *en masse*. Such waste is usually either taken directly to an incinerator or is landfilled, where it should be covered with inert material at the end of the working day. The access to landfill sites should be controlled, thus minimising the opportunity for scavenging by humans and animals. The general public should never be allowed to come into contact with hazardous waste.

For obvious reasons, the workforce responsible for the collection, storage, transport, treatment and disposal of municipal waste is at a greater risk than the general public and special precautions should be taken for their protection. The health of the workforce is safeguarded by the provision of protective clothing, viz. gloves, footwear, face protection.

After a municipal waste landfill has been operating for a time, methane is generated within the waste, which can escape via channels to the landfill surface. The methane can ignite or explode when mixed with air in certain proportions. In a properly maintained landfill, the initial engineering of the site makes allowance for the control of methane but there are redundant sites which were operational before this particular hazard was appreciated (see under Chapter II 5.3, Landfill).

The incineration of municipal waste can result in the production of the toxic gases, hydrogen chloride, hydrogen fluoride and sulfur dioxide. In the proper management of incineration, these gases are trapped and should not enter the atmosphere.

There is an obvious risk in the incineration of hazardous waste. Firstly, it has to be transported to the site and stored prior to incineration. Secondly, the incineration process can emit substances of equal or greater hazard than the starting material, e.g. chlorinated dioxins and chlorinated furans, which are extremely toxic to humans, and which come from the incineration of waste containing PCBs or PCTs. In a properly managed process, the products of incineration are treated safely (see under Chapter II 5.2 Incineration).

The control of hazardous waste is essential to minimise the risk to the health of the handlers. The nature and characteristics of the hazardous ingredients will be known so that proper protection can be provided.

Where hazardous waste might be allowed to be landfilled, e.g. codisposed and diluted with municipal waste, there is an obvious risk to the operator. There have been instances where erroneous mixing of hazardous wastes has produced lethal products, e.g. from wastes containing sulfide and acid.

From the medical view, it is difficult to estimate the risks to human health from the handling of hazardous waste and there have been very few investigations of parameters such as toxicity, infection, flammability and explosivity. There has been little study of occupational mortality and morbidity of the workforce engaged in the waste disposal industry and this is a fruitful area for further research.

The composition of hazardous waste often defies analysis and this is the first difficulty. Even when the identities of the hazardous components are known, there are usually no data concerning interactions between different hazardous substances which may be present in the waste. In addition, the matrices of wastes are variable and there appears to be no way of measuring their effects on the hazardous substances.

3.2 Risk to the environment

In its preamble, the framework EC Directive 75/442/EEC on waste states that all the elements of waste disposal must be carried out without harming the environment. In Article 4, particular reference is made to the parts of the environment which need to be protected from pollution caused by waste disposal. Waste should be disposed of without risk to water, air, soil, plants and animals; generally, it should not cause a noise or odour nuisance or otherwise adversely affect the countryside or places of special interest.

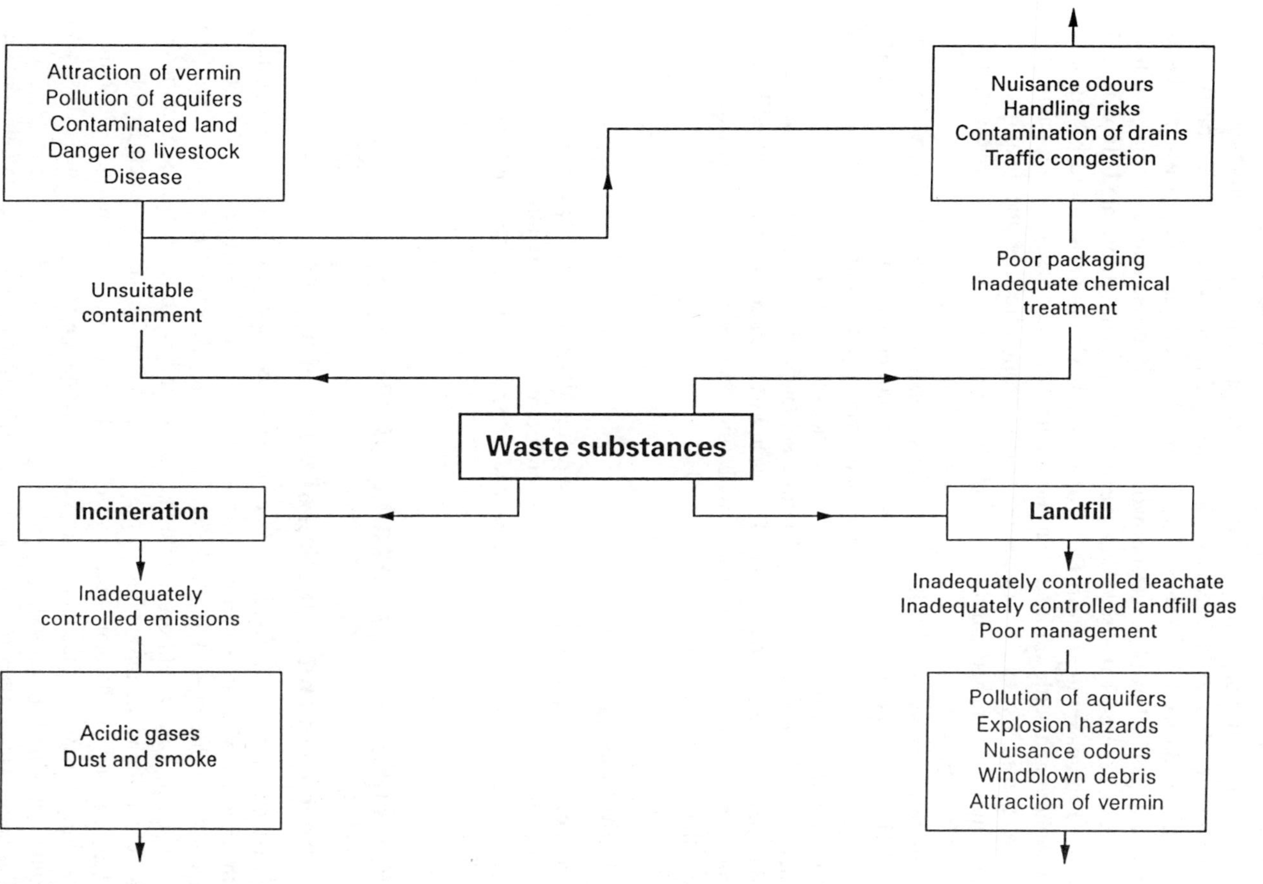

Figure 2 *Hazards of waste to public health and the environment*

In a properly managed disposal facility – landfill or incinerator – there is very little risk that the environment will be polluted.

However, if incineration plant is improperly managed, there is a risk of environmental pollution. The incineration of municipal waste produces acidic gases which, if allowed to escape unchecked, add to the sulfur dioxide burden of uncontrolled industrial emissions. The incineration of hazardous waste produces effluvia which, if allowed to escape, would probably cause similar damage to that by acid rain. There is the risk that the extremely toxic chlorinated dioxins and chlorinated furans emitted from PCB- or PCT-containing waste could be released to atmosphere and settle on land near the incinerator. Any pasture so contaminated could be grazed by cattle, sheep etc. Apart from affecting the health of the animals, these chemicals could be passed along the food chain (see under Section II 5.2 Incineration).

In properly managed landfill waste disposal operations, leachate is contained or treated. However, the escape of leachate is not unknown and nearby watercourses can become contaminated. This can create problems of clean-up where the water is ultimately to be used for human or livestock consumption.

There are occasions when legislation demands the cessation of the use of certain chemicals, e.g. dieldrin, DDT, strychnine. If there is no special collection of these unwanted materials, some will find their way into municipal waste, putting the handlers at risk.

3.3 Conclusions

The collection, transport, treatment, storage and disposal of waste needs control to minimise the risks to humans and the environment. Incineration and landfilling operations can cause continuing risks and these are schematically represented in FIGURE 2.

4 Principles of waste management

4.1 Prevention and reduction of waste at source

The first guideline of European waste management strategy is to prevent waste, but the amount of waste generated throughout the world is increasing.

There is a shortage of landfill sites and some Member States either find the accompanying air pollution problems associated with incineration politically inconvenient or incinerator plant to be too expensive. The EU Council Resolution On Waste Policy, 1990, states that waste should be prevented or reduced at source, particularly by the use of clean or low waste technologies and products; it is now essential that this is practised.

Basically, waste arises at two stages – product manufacture and after the product has been used. Part of Article 3 of the EC Directive on waste, 91/156/EEC, states that Member States shall take appropriate measures to encourage the prevention or reduction of waste production and its harmfulness, particularly by:

- the technical development and marketing of products designed so as to make no contribution or to make the smallest possible contribution, by the nature of their manufacture, use or final disposal, to increasing the amount or harmfulness of waste and pollution hazards
- the development of appropriate techniques for the final disposal of dangerous substances contained in waste destined for recovery

The 1989 Commission's Strategy For Waste Management amplifies this waste prevention policy by advocating the development of clean technologies. It urges industry to reduce the quantity and harmfulness of wastes arising from its production processes and to start this at the design stage. The Strategy also requires waste minimisation in the packaging of commodities. Once packaging is adequate it is environmentally unfriendly to add any more for whatever reason, including sales promotion. The packaging should also be made of re-usable or recyclable materials, in the latter case particularly where the recycling process is carried out in an environmentally benign manner.

4.2 Priority for re-use and recycling

After prevention of waste production and the minimising of waste, the priority in waste management is re-use or recycling.

Recycling was included as a process to be encouraged in the first framework EC Directive on waste, 75/442/EEC. This is reiterated in the Directive, 91/156/EEC, where Member States are commissioned to take appropriate measures to encourage re-use, recycling and reclamation of waste, or any other process with a view to extracting secondary raw materials.

Recycling of waste can take a variety of forms, including regeneration, raw materials recovery and energy conversion. The 1989 Community's Strategy For Waste Management requires that recycling and re-use of waste be vigorously promoted through:

- research and development on techniques
- optimising collection and sorting systems
- reducing the external costs
- creating outlets for the products

Recycling is an ideal and needs to be based on economics and practicality. For instance, quite often recycled paper and plastic items are difficult to sell at an economic price, whereas the markets for materials made from recycled glass and aluminium are usually good. This can be facilitated by suitable tax policies. Before embarking on a recycling programme, it is wise to carry out a life-cycle analysis to ensure economic and environmental viability (see under Chapter III 3.2 Life cycle analysis).

4.3 Proximity, self-sufficiency and collaboration

Before the Basel Convention was approved and despite European Community law regarding the broad principles against the wholesale movement of waste, waste was still moved for final disposal in low-cost facilities, regardless of distance. Provided the running and labour costs were lower than those in developed countries, waste was shipped to developing countries, many of which were not in a position to process it.

The Commission's Community Strategy For Waste Management, 1989, and the Council Resolution on Waste Policy, 1990, called for harmonisation of technical standards for waste disposal plants within the EC. If this could be achieved, waste would only have to be moved to its nearest disposal facility, thus avoiding large movements.

Part of Article 5 of the Directive states that Member States shall take appropriate measures, in co-operation with other Member States where this is necessary or advisable, to establish an integrated and adequate network of disposal installations, taking account of best available techniques not entailing excessive costs. The network must enable the Community as a whole to become self-sufficient in waste disposal and the Member States to move towards that aim individually, taking into account geographical circumstances or the need for specialised installations for certain types of waste. The network must also enable waste to be disposed of in one of the nearest appropriate installations.

The collaboration between Member States is needed to share experience and expertise in the setting up of the network, so that there is self-sufficiency for all. Collaboration is also beneficial at the waste disposal planning stage and in the practice of the safe handling of waste.

5 Public participation

5.1 Education and information

The public needs to be educated regarding the control of waste facilities, including the location of sites in its area. Such information should be available for municipal waste sites via the local government planning authority. In the private sector, responsible industries will provide relevant information concerning their own on-site facilities. Communication with the public needs to be factual and in simple language. Community organisations and action groups are also sources of information, whether this is specific to a local problem or concerning a broad topic, e.g. the pros and cons of incineration versus landfill.

The education of the public is a continuing process, particularly where methods of waste prevention and recycling are concerned. Where communities have bottle, paper, plastic or metal banks, these are increasingly being used.

In schools and in higher education, the application of basic science to environmental problems has helped the younger members of the public to understand the complexities of such topics as safe handling of waste and its disposal. When they become part of the decision-making process, they should have a better understanding than their forebears of the problems associated with waste management.

The public is entitled to the facts concerning the risks to its health and the environment from waste disposal facilities near its homes; also to the arguments concerning planning applications for such facilities.

Since most people learn, initially, through the media about waste management problems, some of the data given may not be adequate. It is essential that the public should have access to facts and unbiased views, but these are not always available. Members of the public can avail themselves of pamphlets and other educational documents from government, industry and action groups. Beyond established facts, there are areas of debate and interested members of the public should be able to take part in these.

Responsible industry has a policy of publicising its objectives and achievements concerning waste disposal, which should include its effective compliance with the legislation and also details of its waste reduction programme.

5.2 Local concerns

The public concern regarding waste disposal is understandable. Poorly managed municipal landfill sites, where, for example, flies, rats and seagulls have been attracted by exposed, odorous, putrescible waste, have given landfill a bad name. Even a well managed site will have frequent vehicle movements in and out of the site, bringing disturbance to a community with traffic and operational noise. If the disposal area has not been filled by the planned time, the life of the site will usually be extended, causing further annoyance to the public. It is not surprising, therefore, that the public has come to mistrust the authorities involved. When a new site is proposed, the normal reaction of the people living nearby is "not in my back yard".

There has been considerable local reaction to having hazardous waste incinerators near or in residential areas. Planning permissions may have been granted on the understanding that the measures taken to control hazardous emissions will always operate efficiently. The nearby residents usually take the view that there is a risk that something nasty will emerge sooner or later. There is also concern about the transportation of the hazardous waste into the site; the last part of the journey will be along the same roads regardless of where the waste has arisen. Local action groups can be formed to represent residents' views either at protest meetings or at planning enquiries. The arguments put forward by action groups have an impact, particularly when cases are put together based on facts and not on speculation and emotion.

Chapter II

Technical aspects

1 Waste management alternatives

1.1 Introduction

The first priority in waste management is the prevention or reduction of all waste, and the second priority is the waste collection. The third priority is the examination of the possible re-use of the waste, either in its found state or in recycled form. Before embarking on a recycling programme, a proper life cycle analysis is needed to ensure that the recycling is a worthwhile option (life cycle analysis is discussed in Chapter III 3.2). The last resorts are treatment and disposal.

As far as industry is concerned, the manufacturing processes need to be studied with a view to minimising the waste. If there is no satisfactory alternative process yielding either re-usable waste or, preferably, no waste at all, then the first option is to examine where waste can be reduced or recycled. This can possibly be achieved at source by using alternative raw materials; similarly, a modification of the manufacturing processes and a critical appraisal of the products can also lead to waste minimisation.

Municipal waste can be separated at source by the consumer into disposable waste and that which can be recycled, e.g. glass and paper. Apart from the simple separation of empty bottles and waste paper, consumers need to be given adequate information on how to separate recyclable materials.

From the public viewpoint, the driving force behind waste minimisation is the achievement of a cleaner environment; this view is shared by industry and commerce, where there could be a resulting reduction in handling hazards and also of production costs.

1.2 Waste reduction

A programme of waste minimisation cannot be achieved without a political will by national governments, including fiscal measures. Waste minimisation is at the heart of any waste management policy and entails a flexible management approach where all facets of the manufacturing processes are continually reviewed. There may be the possibility of changing some or all parts of processes in order to reduce the quantity of waste. This includes a study of the type and quantity of the raw materials used and of possible excesses of packaging of the product.

1.2.1 Programme

A good waste reduction programme should include:

- reduction of the total amount of hazardous waste released to the environment
- development of "waste reduction awareness" in employees
- re-emphasis of the need for continuous improvement in waste reduction
- provision of incentives, and recognition for excellent performance

It is essential to have the commitment of management, who must give the lead in the waste minimisation programme. In so doing, there is an implicit acknowledgement that capital and manpower will be budgeted and available for this purpose. Management should set waste reduction goals, e.g. percentage reduction by a specified time.

Waste reduction has the long-term benefits of creating a cleaner environment and of possibly cutting production costs, which makes good business sense. It could also reduce the long-term environmental liability and handling hazards and, hence, insurance costs of a company.

There has been resistance to the idea of waste reduction, which stems from the natural reticence to change. Some industries have been content to leave the production processes as they are, particularly when there have been no problems and the profit margins have been satisfactory. However, public and media pressures have led to attempts being made to reduce the quantity and toxicity of airborne pollution, as well as of solid and liquid effluent. Environmental legislation to protect the individual and the general environment has helped to focus attention on the problems of waste disposal and this has ensured that responsible companies have looked at waste reduction as a viable option within their waste management programmes.

In some countries, the public is encouraged to reduce domestic waste by the collecting authority restricting the amount or type it will collect on a regular basis. Where it is safe to do so, the householder may be able to re-use items such as cleaned containers which previously had held a purchased product, and so help to reduce waste quantities.

1.2.2 Opportunities

Before investigating the opportunities available to reduce waste, an inventory of wastes arising should be made for every stage in the manufacturing process.

Frequently, raw materials and supplies are bought based on keeping production costs down, without considering waste disposal costs. In order to attempt to reduce the waste and, therefore, disposal costs, it might be possible to:

- reduce to a minimum the number of different associated materials used, e.g. paints, resins and adhesives. This streamlining alleviates shelf-life problems and also reduces the number of partially-used or empty containers which may eventually have to be discarded

- purchase container sizes of perishable items appropriate to the actual use. It can be less expensive to buy litre containers than to buy ten or a hundred litre containers of the item at a lower unit cost, only to have to reject most of the contents at a later date because the expiry date has been exceeded (this applies particularly to small and medium enterprises)
- reduce to a minimum the inventory of hazardous materials and ensure that stock is consumed in chronological order to further reduce waste disposal of out-of-date material
- ensure that machinery is operating efficiently and inspect fluid storage facilities regularly. Leaking tanks, valves or pumps are an obvious cause of needless waste
- exchange or sell the waste to another manufacturer, who can use it as a raw material in his process.

Public awareness about waste reduction has been heightened by witnessing the excessive packaging on many domestic items. The public is also aware of the ever-increasing volume of "junk" mail. Large savings on resources could be made if these areas were looked at from the waste reduction standpoint.

1.2.3 Production Methods

There is a tendency to continue using the same manufacturing processes even though improved methods have been developed. Current technology must always be under review with an eye to improvement in production methods; this will reduce waste as far as practicable and will reduce the environmental impact to a minimum.

1.2.4 Substitution of Materials

The elimination of harmful materials, or their substitution by ones which are more environmentally benign, can help reduce hazardous waste. The reduced cost of disposal or the reduced exposure of the workforce to hazardous material can also justify the change.

1.2.5 Examples

(a)   To reduce hazardous waste, it may be possible to replace hazardous starting materials with safer ones but without significantly altering the efficacy of the product.
      An example of this is to use water-based formulations for paint in place of solvent-based ones. Another example is to change from cadmium or zinc electroplating to aluminium-based coating.

(b)   It has been found that the standard 45-gallon drum can retain more than a litre of liquid when it is up-ended. A redesigning of the shape of the drum has brought this waste volume down to less than 50 ml.

(c)   Since 1975, a company has been engaged in a programme to reduce the need for retrofitted pollution control devices, by minimising pollution at source. This programme has succeeded in eliminating 1.5 billion gallons of industrial wastewater containing about 10,000 tons of water pollutants per annum. This has been achieved by changes in product development, engineering design, product re-formulation and changes in processes.

(d)   Trichloroethylene, used for metal cleaning, can be substituted with a solvent of low toxicity, based on limonene.

1.2.6 Summary

Every item and every step of a manufacturing process should be examined for opportunities to reduce waste. FIGURE 3 is a diagram of a typical chemical manufacturing process, indicating the various points where such reduction is a possibility.

The general public should receive more education, coupled with greater incentives, on how to minimise domestic waste.

1.3 Recycling

Recycling is the collection and separation of materials arising from waste, and subsequent processing to produce marketable products.

After serving its original purpose, recyclable material has, by definition, some value such that, in its remade form, the product has a new purpose and is not a waste. Recyclable material made into a similar product to the original is termed primary recycling (also known as "closed loop"). If the material is made into a different object, this is called secondary recycling.

It should be emphasised that although recycling can lead to waste minimisation, it does not necessarily make the environment cleaner, e.g. if an individual travels by car specially to deliver glass bottles or waste paper to the banks, the net result may be detrimental to the environment. The resource saved by the recycling may be more than offset by the energy consumption, and the accompanying air pollution generated. A proper life cycle analysis should be carried out where there is any doubt as to the efficiency of a waste recycling process (life cycle analysis is discussed in Chapter III 3.2).

1.3.1 General Objectives

Although municipal waste forms only a part of total waste, the bulk being from industrial sources, the recycling of municipal waste has a higher media profile. Considerable research has gone into the successful recycling of paper, cardboard, plastics, glass, aluminium and steel.

There are basic requirements for recycling and these are:

* the collection and the transporting of the waste to be recycled
* the separation and clean-up of the waste
* the processing of the waste to obtain marketable products which have then to be sold

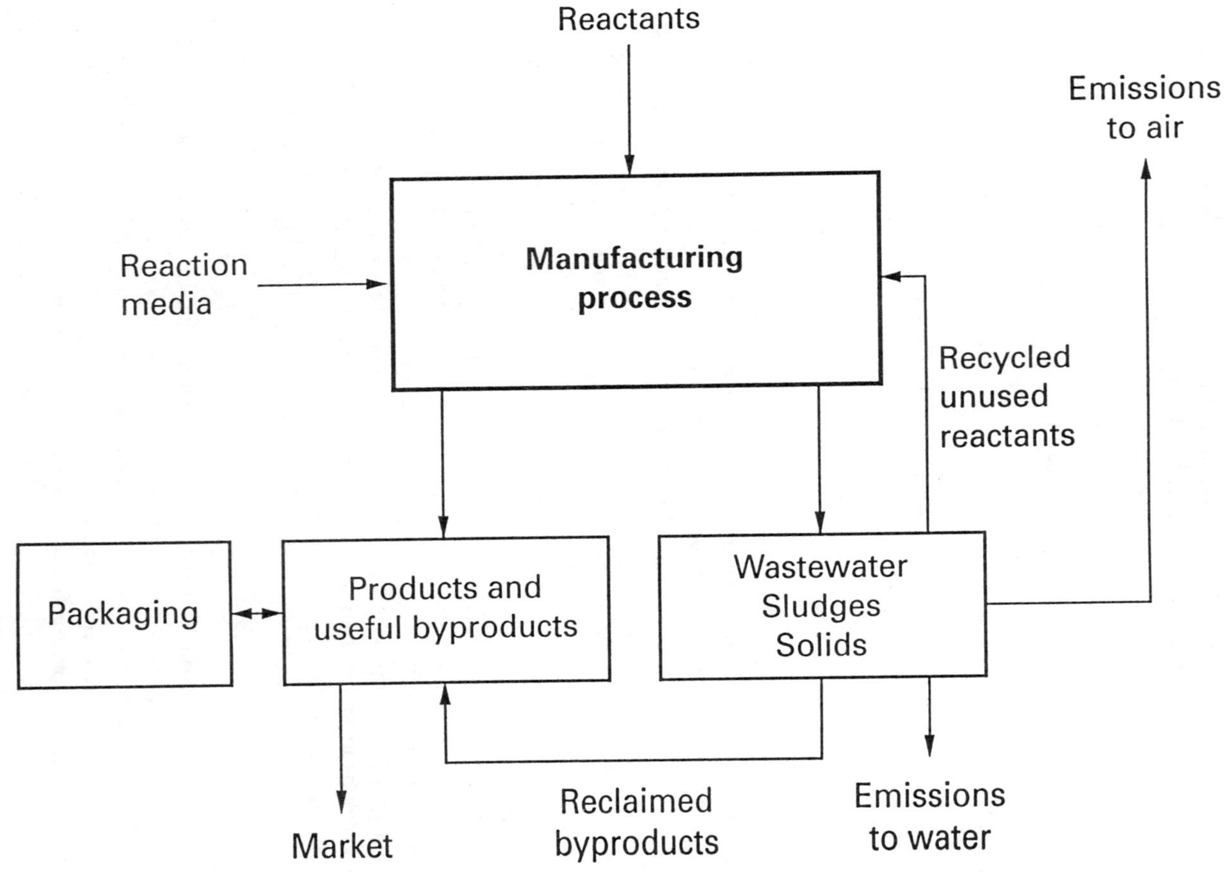

Figure 3 *Chemical manufacturing process indicating waste minimisation*

Types of recycled waste may also be divided into categories according to the part of the manufacturing process from which they have arisen. Process-derived waste can include unused starting materials which can be returned directly to the production process, whereas product-derived waste will, inevitably, have to be reprocessed before being turned into useful commodities. Ancillary waste includes the packaging used for components and products.

The benefits from recycling may include:

* conserving natural resources
* saving energy in production and transport
* reducing the risk of pollution
* saving costs in monitoring
* saving costs in treatment and/or disposal
* reducing the demand for waste disposal facilities and landfill space
* producing goods more cheaply

Public pressure and concern provides an incentive to recycle. Environmental pressure groups have raised the public consciousness about environmental issues and, in turn, the public is applying pressures to various levels of government. There are various ways of encouraging recycling and these include:

* providing funding to assist technological advances and improved design (to make products more recyclable)
* adopting a policy of discrimination in favour of recycled products
* encouraging industry to reduce unnecessary packaging of consumer goods and to take a measure of responsibility for dealing with waste from packaging (ideally, the packaging should be recyclable)
* educating the public in the merits of recycling

1.3.2 Marketability and Economics

Before embarking on a recycling programme, a secure and stable market for the recycled products should be established. This part of the recycling process is often more difficult to achieve than the technology originally required to obtain the product. The recycled material will have to yield a consistent product in terms of quantity and quality. I.e. there must be enough to fulfil the marketing obligation and any contaminants must be removed to meet the specification required by the purchaser.

The need to have a constant supply of recycled material has proved difficult in some instances in the past and, nowadays, the industrial purchaser will probably require a guarantee that the supply will not dry up. Virgin material may have to be added to the recycled product on occasions to make up the shortfall, and this will have a bearing on the total cost.

The success of recycling schemes depends on sound marketing. For recycled material to compete with virgin material, it has to be sold at a comparable price. However, there are those who believe that recycling is an environmental ideal

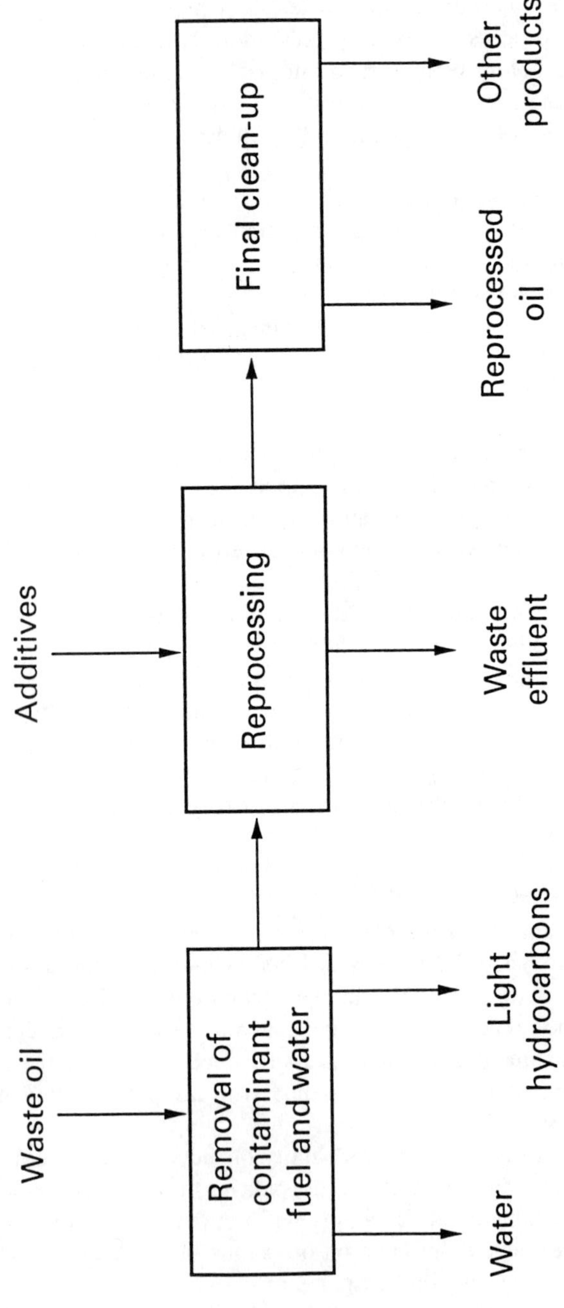

Figure 4 *Waste oil processing*

function and that should the process not be economically viable, the cost incurred is justified if this means a cleaner environment.

An industrialist would not have the collection and transport costs if the material is recycled within the confines of the site. Also, the industrialist may have more control over the quality of the material to be recycled and would be in a position to keep contamination to a minimum; this would avoid excessive clean-up costs. In contrast, the municipal waste for recycling will have collection and transport costs, and clean-up before re-manufacture may be expensive.

The recycling of glass, plastics, aluminium and steel by the industry which made these packaging materials in the first place ought to be economically viable. The economics of the recycling of paper and cardboard, particularly from municipal sources, is more uncertain because of large variations in market demand. Depending on the market, the selling price for recycled paper products may be too low to make the recycling worthwhile.

The main benefit to all concerned is in the savings on treatment and/or disposal costs. In those industries where hazardous waste is a by-product, it makes sound economic sense to segregate non-hazardous from hazardous waste efficiently, since there is a chance that waste might be recyclable. Should this waste become contaminated with a hazardous material, it will have to be treated as hazardous, which means that treatment and/or disposal will increase the overall costs.

1.3.3 Examples

(a) Defunct lead-acid car batteries are recycled to recover lead and also plastic (from the casing). The lead can usually be processed into other car batteries. The recovered plastic is normally reused as a lining material and not for battery casing.

(b) If materials recovery is not practicable and the waste is combustible, it may be usable as a fuel. Using waste as an energy source has much to commend it, provided the enabling technology has been achieved. The efficient incineration of municipal waste is a good example of this, as is the use of waste oil as a fuel.

(c) Waste oil is also reprocessed for use as a lower grade product, usually as a lubricant ingredient. With municipally-collected oil, the extent of contamination is unknown and industrially-collected oil will be contaminated with a variety of impurities. FIGURE 4 is a diagram showing the various stages of clean-up; water, unwanted organics and metals are removed in the process.

(d) Improved technology has meant that the segregation of materials has become easier. It is possible to break up a motor car for re-use and to separate the glass, plastics and ferrous and non-ferrous metals. Some motor cars are being designed with such disassembly in mind.

2 Waste handling, transport and storage

2.1 Identification, characterisation and analysis of wastes

Waste can be a single substance with a specific name; it is more likely to be a complex mixture with no name other than a description of its physical type, e.g. liquid, sludge, slag, semi-solid, solid, and accompanied with a qualification which describes the process or part of the process from which it has been rejected, e.g. metallic, organic, oily. Other descriptions of waste rely on the type of matrix or industry from which it was derived, e.g. domestic, agricultural, industrial.

The identification of waste relies, to a large extent, on a knowledge of its derivation and history. Precise chemical and/or physical analysis of waste can be extremely difficult and, at times, impossible, and is usually only carried out as a last resort, or for quality assurance purposes if the ingredients are already known.

The characterisation of waste is a more meaningful concept than identification. The absolute identification of ingredients may not be required, provided the character and properties of the waste are known. Furthermore, it is often only necessary to establish the nature of the waste, and if it proves to be relatively innocuous, its precise composition need not be an important factor.

For all practical handling purposes, there are only two types of waste – non-hazardous and hazardous. By general consensus, non-hazardous waste is taken to be municipal (household, domestic) waste, together with commercial and some industrial waste. The waste on which much attention has been focussed is that which is hazardous *per se*, or which contains hazardous substances in sufficient concentration to make the waste hazardous. The difficulty of applying rigid rules to the demarcation between non- hazardous and hazardous waste is recognised in the complex system of listing in the EC Directive 91/689/EEC, Article 1.4. Properties which render a waste hazardous are listed in the Directive as explosive, oxidising, flammable, highly flammable, irritant, harmful, toxic, carcinogenic, corrosive, infectious, teratogenic, mutagenic and ecotoxic. Substances and preparations which yield toxic agents when in contact with water, and also leachates are included in this list.

In general, the criteria for identifying the extent of the hazardous character of waste include the following:

- the knowledge that there are one or more hazardous substances present
- the concentration or chemical reactivity of the hazardous substances
- the physical form of the hazardous substances
- the mobility and persistence of the hazardous substances

Work commissioned by WHO has resulted in a list of 21 chemicals with a high hazard rating. All of these chemicals could be present in waste presented for disposal at landfill sites. TABLE 6 lists these chemicals, the presence of which would confer hazardous status on the waste, depending to some extent on the concentrations of the chemicals in the waste. The hazard rating of each chemical was evaluated according to a procedure which took into account the concentration of the chemical, its environmental fate and transport, and its effect on human health. The chemicals are divided into two "divisions", the first 12 being judged as more hazardous than the second 9. The limitations of this approach were the use of value judgements and the unfortunate absence of data on waste characteristics and practices; nevertheless, it is a starting point from which other lists may follow.

Table 6 *List of priority chemicals in hazardous waste (work commissioned by WHO)*

1,2-Dichloroethane
Benzo [a] pyrene
Chloroform
Carbon tetrachloride
Benzene
PCBs/PCTs
Trichloroethylene
Chromium (VI)
Tetrachloroethylene
Monochloroethylene (vinyl chloride)
Mercury
Cadmium

Dichloromethane
1,1,1-Trichloroethane
2,3,7,8-Tetrachlorodibenzo-p-dioxin
Lead
Asbestos
Arsenic
Phenol
Methyl ethyl ketone
Cyanide

Where wastes are of unknown origin, although being suspected of containing hazardous material, it is sometimes possible to achieve the identification and quantification of the hazardous material by physical and chemical analysis. A physical separation of components of heterogeneous waste can be the prelude to the analytical determination of each component. Basic textbook chemistry is all that is required at this point, i.e. division into inorganic and organic parts. Inorganic wastes typically can include inorganic acids, alkalis, cyanides, sulfides and metals. Organic wastes can be divided into biodegradable and persistent wastes; most food components, for example, are biodegradable whereas the classic examples of persistent organics in waste are those containing PCBs/PCTs and organochlorine pesticides.

Detailed chemical analysis, whether inorganic or organic, is carried out using, where possible, proven and accredited qualitative and quantitative analytical methods. This can require expensive instrumentation and consummate skill, and should only be undertaken by suitably trained personnel. Methods of extreme sensitivity are called for when seeking trace amounts of the more hazardous chemicals, e.g. lead, mercury, PCBs/PCTs and organochlorine pesticides. However, resorting to such in-depth analysis may only be necessary in special cases and the bulk of analytical work is in establishing the characteristics of the waste components, e.g. the determination of pH, flash point, total solids, conductivity, total ash, acid- and water-soluble ash, BOD and COD.

2.2 Safe practices in collection, handling and transport

As part of a proper waste management programme, waste should be collected and handled in a professional manner, which includes handling the wastes safely. Waste should be handled by properly trained personnel and every transaction (except for municipal waste) should be documented in such a way to give maximum information on the type of waste, its quantity, its history including all the manufacturing detail, its intended fate and, not least, its hazardous nature. As a sensible precaution, waste of unknown origin or history should be treated as if it was hazardous.

The collection of waste takes a fraction of the time when compared to the length of time that waste is sometimes stored, albeit temporarily. However, storage time should be kept to a minimum for both environmental and economic reasons. Factors which should be considered when storing waste include packaging, compatibility, segregation, ventilation, the environment and regulatory compliance. An appraisal of these factors is essential for the storage of hazardous waste, which should be treated with the same respect as the hazardous chemicals or substances that it contains.

2.2.1 Waste containers

Waste containers should not break, wear out or be subject to corrosion; they should be designed to prevent leakage and accidental spillage and be secure enough to prevent vandalism by humans or animals. Care should be exercised where low density waste, e.g. paper, is handled, to prevent it being blown away. It is imperative that containers are accurately labelled with all relevant information and the labelling itself should be secure and the writing indelible.

Hazardous waste containers should be made of a material which is non-reactive with the waste they contain or the hazardous constituents of that waste. Plastic is relatively non-reactive compared to metal. Wastes including "sharps" must be contained in such a manner that there is no possibility of injury to handlers.

Temporary containment of non-hazardous or non-putrescible wastes can be in open containers such as skips.

2.2.2 Compatibility and segregation

If for reasons of economy of space, different wastes are stored together in the same containers, care must be taken to ensure their compatibility, e.g. sulfide- or cyanide-containing waste must never be stored with acidic waste because of the almost certain evolution of hydrogen sulfide or hydrogen cyanide as a result.

Segregation of waste is a consideration when the viability of subsequent disposal methods could be jeopardised, either for technical or economic reasons. As an example, the mixing of waste chlorinated organic solvents and similar non-chlorinated solvents is not done where fuel recovery is planned from the non-chlorinated solvents.

2.2.3 Ventilation and the environment

However sound the containment of waste containing highly volatile substances, it is sensible to store containers in a well ventilated area. Escaping vapours can cause adverse health effects and also there would be a possible risk of fire or explosion. Similarly, the general environment in which containers are stored should be such that there are no excesses of heat, cold, moisture or light.

2.2.4 Regulatory compliance

National regulations may be in force to determine the conditions of containment and storage and these must be followed. Where regulations are in force, it will be found that the general principles outlined above are normally included and are concerned with the receiver of the waste as well as the producer and handler.

2.2.5 Waste transfer

Having properly contained, labelled and stored the waste, its transfer should be carried out in a manner which protects the health of all concerned in the transfer. Any national law in force will usually require extensive documentation. This will include a complete record of the waste itself (as for storage), companies or personnel involved and dates of production, storage etc.

The owner of the waste must not deposit, dispose of or treat it in an unauthorised manner. The transfer should only be made to an authorised receiver, who will also receive the full description of the waste from the owner; this is particularly important in the case of hazardous waste. Adopting the principle that hazardous waste should be treated as if it was wholly comprised of hazardous ingredients, a knowledge of the nature and characteristics of the hazardous ingredients is essential. The documentation accompanying the transfer of hazardous waste should contain details of the hazards of the ingredients in terms of oral and inhalational toxicity (essential when asbestos or finely divided metal oxides are known to be present), dermal toxicity, flash point, explosivity, and physical characteristics such as vapour pressure and boiling point, where applicable. Unless toxicity tests have been carried out on the waste itself, acute hazards posed by the waste can only be estimated by a knowledge of the hazards of its components.

3 Transboundary movement of waste

3.1 The Basel Convention

During the 1980s a number of national, regional and international initiatives were taken to address the problems caused by the growing international trade in hazardous waste transfer. In 1989, the fourth Lomé Convention of the EEC and African, Caribbean and Pacific countries specified that hazardous waste would not be exported by the EEC to those countries. This strategy, in a broadened form, is the cornerstone of the Basel Convention. The Basel Convention is the international agreement reached, through UNEP, for the control of transboundary movements of hazardous wastes and their disposal. In March 1989, 116 countries plus the EEC adopted the Convention.

At this time, 300–400 million tonnes of hazardous waste were estimated to be generated throughout the world. This had resulted in increasing stockpiling of wastes containing such items as corrosive acids, organic chemicals and toxic metals, which pose serious long-term health and ecological threats.

The transboundary movement of waste which usually causes concern is that from an industrialised country to a developing one. This results in problems for the receiver, who might lack the expertise and resources to deal with the waste in an environmentally sound manner.

The Basel Convention consists of 29 Articles and 6 Annexes.

3.1.1 Main Provisions

The main provisions, briefly, are as follows:

Overall aim
The reduction of all waste and the discouragement of transboundary movements.

Disposal
The disposal of waste in an environmentally sound manner as close as possible to its source.

Export
Hazardous waste to be exported only if the waste generator lacks the resources or expertise to ensure its correct disposal.

Import
Importing State must possess the proper resources/expertise to fulfil the above obligations.

General
(a) States which are party to the Convention are required to limit any export/ import arrangements with those outside the Convention, to those which can demonstrate that they possess environmentally sound waste disposal procedures.

(b) Transboundary waste movements must be carried out according to control measures stipulated by the Convention.

(c) Waste for transboundary movement must be packaged, labelled and transported in conformity with internationally recognised and approved practices.

(d) Co-operation, technical assistance, and information transfer on waste management and disposal, is to be actively encouraged between States within and outside the Convention.

3.1.2 Convention definitions of hazardous and other waste

The Convention defines waste as "substances which are disposed of or are intended to be disposed of or are required to be disposed of by the provision of national law". TABLE 7 is a summary of Annex IV of the Convention, which lists the practical disposal operations, whether or not they lead to the possibility of resource recovery, recycling, reclamation, direct re-use or alternative uses.

Table 7   *Disposal operations as outlined in Annex IV of the Basel Convention*

Operations not leading to possibility of resource recovery, recycling, reclamation, direct re-use or alternative uses

Deposit onto or into land (e.g. landfill)
Land treatment (e.g. biodegradation of liquid discards in soil)
Deep injection (e.g. pumpable discards into wells)
Surface impoundment (e.g. placement of liquid discards into lagoons)
Specially engineered landfill (e.g. placement into lined discrete cells)
Release into water (either river, lake, sea)
Biological treatment
Incineration on land or at sea; (except for energy recovery)
Permanent storage (e.g. emplacement of containers in a mine)

Operations which may lead to resource recovery, recycling, reclamation, direct re-use or alternatives uses

Use as a fuel or other means to generate energy
Solvent reclamation/regeneration
Recycling/reclamation of organics other than solvents
Recycling of metals, metal compounds and other inorganic materials
Regeneration of acids or bases
Recovery of components used for pollution abatement and from catalysts
Re-refining of used oil
Land treatment benefiting agriculture or ecology

The Convention differentiates two types of waste – hazardous and "other". Hazardous waste is defined by reference to Annexes I and III of the Convention; Annex I lists 18 different waste streams and 27 chemicals which could be present in waste – all these are to be taken as hazardous. Annex III contains a list of 13 hazard characteristics, which formed the basis of the selection of the substances in Annex I. If a waste stream listed in Annex I is found not to possess any of the hazard characteristics of Annex III, it is considered to be non-hazardous. Overriding these considerations, national legislation may deem a waste to be hazardous which is not represented by either Annexes I or III. The hazard characteristics listed in Annex III are shown in TABLE 8.

"Other wastes" are listed in Annex II and are household wastes and incinerator ash. (Radioactive waste is not covered by the Convention).

Table 8 *Summary of hazardous characteristics of waste as given in Annex III of the Basel Convention*

Explosive
Flammable (liquids)
Flammable (solids)
Liability to spontaneous combustion
Liability to emit toxic gases after water or air contact
Oxidising
Organic peroxide oxidising
Poisonous (acute)*
Infectious
Corrosive*
Liability to emit flammable gases after water contact
Toxic (delayed or chronic)*
Ecotoxic*

*Criteria laid down in EC Directive 67/548/EEC

3.1.3 Control measures and illegal traffic

Where transboundary movements are generally allowed, there is an elaborate control system which is based on consent following the provision of information. The rights and obligations of the states for the export, import and transit of the waste are laid down. For instance, the exporter must specify the reason for the export and the names of the exporter, generator, the site of generation and the process by which the wastes are generated. The nature of the wastes and their packaging, as well as the intended itinerary, must be specified, together with the location of the disposal site, the name of the disposal contractor and the method of disposal. There are a number of administrative procedures to be carried out, some of these involving the transit States, i.e. the States through which the wastes may pass before reaching their destination.

If all the control measures are not carried out and the waste is still exported, this constitutes illegal trafficking, which is a criminal offence; the receiver is then under an obligation to ensure that the waste is returned to its origin in a safe manner.

3.2 EU Council Regulation (EEC) No. 259/93

3.2.1. Introduction

In 1984, the EC produced its first Directive, 84/631/EEC, on transboundary movement of hazardous waste. The Directive defined a notification system for waste movement, starting from the generation and ending with the disposal. It was based on the obligation of prior detailed information provided by the exporter. The Directive was also intended to allow the Community to have data relating to the nature of waste and its characteristics, and to inform the Member States concerned of the planned transfers while giving them the opportunity to object. Before the enforcement of this Directive, transboundary waste movements were not controlled in a legal manner, and even when the intended control was in place, the supervision was lax. This Directive was modified in 1986 (86/631/EEC) and this introduced additional provisions to tighten control. It also took into account developments taking place within both OECD and UNEP, which greatly concerned the control and safety of shipments to developing countries.

EU Council Regulation 259/93/EEC was adopted on 1 Feb 1993 and deals with the supervision and control of shipments of waste within, into and out of the European Community. Apart from the two previous Directives mentioned, the precursor of the Regulation was an OECD decision concerning the control measures for transboundary movements of wastes intended for recovery. The OECD decision also defined three categories of waste: the green list of non-hazardous or "green waste"; the amber list of hazardous waste and the red list of very hazardous waste.

Following adoption of the Basel Convention by the EEC, the OECD decision formed the basis of the Council Regulation, the primary objective of which is to improve the system of control and supervision of all waste transfers within the Community. Formal approval of the Basel Convention by the EEC is given in 93/98/EEC.

3.2.2. Main provisions

The main provisions of the Regulation cover notification procedures and the details required for the consignment note.

The Regulation contains the green, amber and red lists as outlined above as Annexes II, III, and IV respectively. Waste in the green list is subject to a simple trade control in its transboundary movement, with certain information having to be recorded; waste in the amber list is subject to a procedure of written notification (prior informed consent) in advance of the transboundary movement; waste in the red list is controlled in a more strict way than waste in the amber list, the importer or transit state having to give prior written consent before the start of the transboundary movement. In October 1994, a Commission Decision (94/721/EC) modifying the green, amber and red lists was adopted. Table 9 is a résumé of the three lists.

Table 9 *Lists of wastes from EC Decision 94/721/EEC*

Red list	Amber list	Green list
PCBs/PCTs at >50 mg/kg	Metal bearing wastes	Metal and metal alloy wastes
Tarry residues from refining, distillation and other pyrolytic treatment (excluding asphalt cements)	Principally inorganic constituent wastes which may contain metals and organic materials	Metal-bearing wastes from melting, smelting, refining
Asbestos dust and fibres	Principally organic constituent wastes which may contain metals and inorganic materials	Other non-hazardous wastes containing metals
Ceramic-based fibres, similar to asbestos	Wastes which may contain either inorganic or organic constituents	Wastes from mining
Congeners of polychlorinated dibenzo-furan		Glass wastes
Congeners of polychlorinated dibenzo-dioxin		Ceramic wastes
Leaded anti-knock compound sludges		Other non-hazardous wastes, principally of inorganic constituents, which may contain metals and organic materials
Peroxides (excluding hydrogen peroxide)		Plastic wastes
		Paper, paperboard and paper product wastes
		Textile wastes
		Rubber wastes
		Untreated cork and wood wastes
		Wastes from agro-food industries
		Wastes from tanning, fellmongery and allied industries
		Other non-hazardous wastes, principally of organic constituents, which may contain metals and inorganic materials

The notifier who ships the waste for disposal has to notify the competent authority of destination and any transit State body involved. The consignment note must contain details concerning the waste, and must include source, composition, quantity, the name of the producer, the name of the consignee, the location of the disposal centre (which must have adequate technical capacity for the disposal of the waste under conditions presenting no danger to human health or to the environment), and arrangements for routing and insurance. There is an acknowledgement procedure for the authority of destination, and various lengths of time to complete the authorization procedure.

Where waste is being shipped for recovery, which may be relevant to waste on the amber list, a similar notification procedure applies. The location of the recycling centre is a requirement on the consignment note. This must contain the method of disposal for the residual waste after recycling has taken place and also the amount of recycled material in relation to its residual waste.

The export of waste for disposal outside Member States is prohibited except to EFTA countries which are party to the Basel Convention. The export of hazardous (Amber and Red Lists) waste for recovery outside Member States is prohibited except to those countries to which the OECD decision or the Basel Convention applies or where bilateral or multilateral agreements apply. Commission Decision 94/575 controls certain shipments of waste to certain non-OECD countries.

The import of waste into the Community is prohibited except from EFTA countries or countries party to the Basel Convention, or where there is bilateral or multilateral agreement.

The notification procedures for the export and import of wastes involving non-Community countries is similar to that for Community countries; the same procedures also apply to any transit countries involved. All exports of hazardous waste to African, Caribbean and Pacific countries are prohibited.

3.3 Regional Collaboration

Legal requirements apart, where more than one country or State is involved in a transaction such as transboundary movement of waste, collaboration between the exporting and importing parties is essential. This has been formally recognised by the Basel Convention in its seeking of international co-operation, technical assistance, and transmission of information in areas related to environmentally sound waste management. Knowledge of development of low waste technologies, training for appropriate personnel, and data concerning the effects of waste management on human health and the environment should be passed on to others, particularly the developing countries. Under the Convention, developing countries are entitled to receive all possible help from developed countries.

Collaboration between states on the harmonisation of technical standards and guidelines is also encouraged by the Basel Convention.

An example of a working collaboration is that between Commonwealth countries. The Royal Society of Chemistry co-operates with the Commonwealth Science Council on waste management. Centres have been set up in the West Indies, India, Kenya and Nigeria with the purpose of improving exchanges of information on a regular basis so that collaboration can take place regionally and globally on waste management. Scientific meetings have been held in these centres as a contribution to the Commonwealth Science Council's training programme.

4 Waste treatment

4.1 Introduction

Treatment can only usually be applied to liquid waste. Treatment is used either to degrade hazardous waste completely or, more usually, to modify its physical, chemical or toxicological properties to yield a waste which can be disposed of safely.

The kind of treatment necessary depends on the physical form and hazardous nature of the waste. Detoxification may be achieved by chemical treatments such as neutralisation or oxidation/reduction. Biological treatment, such as aerobic or anaerobic destruction is being increasingly utilised. There are technologies aimed at reducing volume, particularly for liquid wastes, e.g. precipitation, dewatering and phase separation; and for solid, less hazardous municipal waste there is compaction. Hazardous components can be immobilised by solidification processes.

Where the hazardous waste production is insufficient to warrant an on-site disposal unit, wastes may be handled by a specialist contractor who will gather similar wastes from more than one source so as to make his operation economically viable.

In practical terms, the treatment technologies are used sequentially according to cost effectiveness. Physical, chemical and biological processes are employed as necessary. Basic treatments such as settlement and neutralisation are followed by intermediate treatments, including metal precipitation and biodegradation. Any final advanced treatments are usually ones of refinement, and include ion exchange, reverse osmosis, adsorption and solvent extraction.

4.2 Physical treatment

Physical treatments can be divided into the more simple initial treatments, such as screening, sedimentation, filtration and decanting; and the more sophisticated treatments, such as adsorption and molecular separation. TABLE 10 summarises the various physical methods available for waste treatment together with examples of their uses.

Table 10 *Physical methods of waste treatment*

Process	Example
Screening	Removal of coarse paper fibres
Centrifugation	Sludge dewatering
Filtration	Sludge dewatering
Sedimentation	Removal of settled solids in coal production
Dissolved air flotation	Removal of oil and grease
Reverse osmosis	Purification of electroplating rinse solutions
Adsorption	Removal of dyes
Distillation	Recovery of solvents and oil
Solvent extraction	Extraction of PCBs from transformer oils

4.2.1 Sedimentation, flocculation, flotation and filtration

A vital first step in the treatment of wastewater is the removal of suspended solids. The experience gained in techniques historically applied to municipal wastewater is exercised for the treatment of hazardous waste. Sedimentation, i.e. the leaving of a liquid to stand until solid material settles out, is a simple, cost-effective technique, which is widely used for so-called dewatering. Sedimentation may be achievable by coagulation for colloidal solutions, i.e. by using a flocculating agent such as polyelectrolyte or an aluminium salt. When settlement under gravity is not achievable within a reasonable time, e.g. for fine particles, the liquid can be centrifuged.

Suspended organic matter can sometimes be brought to the liquid surface by flotation techniques such as dissolved air flotation. Air is dissolved in the suspension under pressure and minute air bubbles attach themselves to the particles causing them to float to the surface; the particulate can then be skimmed off. This technique is used in petrol refining.

Filtration, using paper media or natural products, such as diatomaceous earth, (also, formerly, asbestos) is an established technique for separating solid from liquid. The technology has advanced to the use of hyper- and ultra-filtration, where very finely divided solids can be filtered. Aqueous suspensions from oil industrial waste are ultra-filtered to remove suspended oil and grease droplets and, in another example, following the washing of paint-making tanks, paint residues from the resulting suspension are ultra-filtered; the residues are subsequently recycled and the water is returned for re-use.

4.2.2 Reverse osmosis

Reverse osmosis is a membrane separation technique where the membrane is selectively permeable to water but excludes ionic solutes or emulsified oil. The separation of oil from oil/water emulsions used for metal cutting is an example of this.

4.2.3 Adsorption

Hazardous waste organic materials can be removed from wastewater by adsorption on to activated carbon. This is a particularly useful technique where the contamination is by organic substances of high molecular weight, e.g. dyestuffs and some pesticides. Because of the relatively high cost of the activated carbon, the technique is often restricted to a refinement of the wastewater clean-up.

4.2.4 Distillation

Distillation is used most often for the separation of solvents. This technique is favoured in the oil industry where the recovery of the oil as well as the solvent means that both components of the waste stream can be re-used or recycled. In the manufacture of penicillin, the waste stream is distilled to recover butyl acetate.

4.2.5 Solvent extraction

The most frequently used example of this technique is the use of an organic solvent to extract an organic component from an aqueous medium. This technique is used in the coal coking and also petrol refining processes where phenol needs to be removed from the by-product water. *N,N*-dimethylformamide is used to remove PCBs from waste transformer oils.

4.3 Chemical treatment

If practicable, the chemical treatment of hazardous waste should result in a complete breakdown of the hazardous constituents into re-usable or easily disposable products of low toxicity. In practice, the usual result is a modification of the hazardous nature of the waste; for example, the precipitation and collection of the relatively small amount of the hazardous solid, or by neutralising excess acidity or alkalinity.

4.3.1 Acid/Base neutralisation

Many hazardous wastes are corrosive. Corrosivity is frequently a function of pH. Adjusting the acidity or alkalinity of a waste to create a neutral product is the best means of removing the corrosivity. The product is not only rendered much less harmful, but may often be used in another process. An example of this is in the production of titanium dioxide where there is a sulfuric acid-containing discharge. Lime or limestone is used for the neutralisation of the waste stream and the resulting gypsum may be subsequently used as a soil conditioner.

4.3.2 Chemical precipitation

Chemical precipitation methods are used for the removal of heavy metals, e.g. copper, nickel, zinc and chromium, from electroplating bath effluents. The addition of slaked lime or sodium hydroxide precipitates water-insoluble hydroxides which are filtered off. Other precipitants used for the removal of heavy metals include sodium sulfide and thiourea, both of which yield water-insoluble sulfides. Sulfide precipitation is also used as a "polisher" following the initial more crude hydroxide precipitation.

4.3.3 Oxidation/Reduction

Leaving waste to become oxidised in air usually results in a less hazardous oxidised product. However, the process is very slow and oxygen treatment or, better, ozone with oxygen treatment can be an efficient waste destroyer. Industrial wastewaters containing phenols, aldehydes, cyanide, nitrite and ferrous iron can be oxidised in this way. The treatment of swimming pool water illustrates the use of either ozone or chlorine in the form of hypochlorite as efficient oxidising agents for unwanted organics in wastewater.

Chlorine or hypochlorite is the oxidant used for the oxidation of cyanide. Waste cyanide-based heat treatment salts can contain between 2 and 40% sodium cyanide, and the waste salts arise as solidified melts, which present a handling problem. The cyanide is dissolved in water and then treatment oxidises it to the much less toxic sodium cyanate, with the possibility of further oxidation to nitrogen and carbon dioxide. Cyanide present in the rinse waters from the electroplating industry can also be oxidised in this way.

Reduction does not have the same widespread use as oxidation. The classic example of reduction is in the chromium plating and surface metal treatment industries, where waste solutions of toxic chromates are treated with sodium metabisulfite to yield much less toxic chromium salts.

4.3.4 Electrolysis and ion exchange

Electrolysis is most widely used in the recovery of metals from electroplating effluents. Metals can also be recovered electrolytically from the wastewaters and rinse waters originating from metal finishing operations and from the electronics industry. The metals recovered in this way include cadmium, copper, zinc and lead.

Low concentrations of heavy metals in solution, e.g., copper, can be removed by cation-exchange resin treatment; similarly, low concentrations of anionic species, e.g. nickelocyanide and chromate, can be removed by using cation-exchange resins.

4.4 Biological treatment

The aim with biological treatment is biodegradation, i.e. the conversion, by biological process, of the original organic material into simple inorganic species.

Many different types of industrial waste are treated by biological methods which are similar to those used for the treatment of sewage. Because many hazardous wastes contain toxic substances, there is the possibility that these will kill the micro-organisms necessary to perform the degradation. Provided the concentrations of toxic substances are kept low by dilution of the waste, the micro-organisms usually survive. Degradation of organic substances mixed with topsoil, as in land farming, is due to the action of micro-organisms, e.g. bacteria, present in the soil.

Some bacteria function best in aerobic conditions; others in anaerobic conditions. The common example of aerobic biological treatment is in the activated sludge process for the clean-up of sewage and also of landfill leachates. Chemicals such as acetaldehyde and aniline are more efficiently biodegraded under anaerobic conditions. The use of enzymes in biological treatment of waste is made where there are specific chemicals to be destroyed; for example, the very toxic organophosphorus insecticide, parathion, can be enzymatically converted to the much less toxic *p*-nitrophenol.

4.4.1 Land farming and composting

Land farming is the mixing of solid hazardous waste with topsoil so that the natural soil-containing bacteria can degrade the hazardous waste constituents. The greatest use of land farming is in the biodegradation of waste from the petrol-refining industry.

Composting is the microbiological action of bacteria on waste itself without necessarily mixing it with any other material. Composting is popularly used for the degradation of vegetation: it is also a common method of waste treatment.

5 Waste disposal

5.1 Introduction

The most common waste disposal processes for solid wastes are incineration and landfill. There can be no way of disposing of waste completely since there will always be by-products of a disposal process, i.e. waste from waste.

Economics play an important part in the choice of the waste disposal method. The costs of maintaining incineration equipment are considerably higher than those of landfill. However, as more incinerators are being commissioned, and as landfill sites, already at a premium, become scarce, the cost differential between the two methods is expected to narrow.

5.2 Incineration

Incineration is the controlled combustion of waste originating from municipal, commercial and industrial sources in order to destroy it or transform it into less hazardous material. Organic constituents break down to give carbon dioxide, steam and other gases, such as sulfur dioxide and nitrogen oxides; inorganic constituents are dehydrated and calcined.

In the incineration of municipal and typical commercial waste, the main objective is to reduce the volume and so save considerably on the transport costs associated with landfill and extending the life of the landfill, the finding of which is at a premium.

In the incineration of industrial waste, the objectives are to destroy any hazardous or infectious organic materials present and also to reduce volume. There are incinerators specially designed to cope with the destruction of "difficult waste", i.e. waste which contains hazardous materials, and these incinerators are usually operated on a service or "merchant" basis, i.e. incinerators operated by private companies, where similarly contaminated wastes may be batched. A company which produces large quantities of combustible waste may find it more economical to have its own on-site incinerator. Hospitals may have on-site incinerators for destroying clinical waste.

Properly managed incineration will reduce the volume of the initial waste, but there will still be gaseous emissions and an inorganic residue left at the end of the process. Although hazardous organic materials should be destroyed in the process, the emissions and residues contain pollutants which have to be contained and disposed of in due course (see Chapter II 5.4 Waste from waste).

Municipal and typical commercial waste is incinerated at a temperature of up to 850 °C in the presence of air (or oxygen); most industrial waste incineration is carried out at higher temperatures, i.e. 1200–1250 °C, to ensure the complete destruction of hazardous organic materials. For efficient incineration, not only the temperature but also the "residence time" of the waste is important. The physical problem of attainment of enough turbulence for the mixing of the air with the burning waste is a major factor.

A flow diagram of the basic incineration process is shown in FIGURE 5. This shows that solid and liquid waste may be mixed together after an initial preparation stage; solids can be screened and broken up; liquids can be filtered and settled.

The most efficient mixture for incineration is deduced from a knowledge of the combustibility of the components, and it is possible to mix aqueous solutions with miscible organic waste to give a combustible product. Otherwise, fuel, often in the form of waste oil, can be added to ensure complete combustion. Air (or oxygen) is always required for incineration. If air is omitted, waste is pyrolysed to produce a fuel which is then burned to produce energy.

Gases from incineration have to be cooled and then cleaned of particulate matter before removal of air pollutants. The latter is achieved by a combination of filtration, precipitation and scrubbing techniques. Some particulate matter, e.g. fly ash and dust, will be deposited in the flue and on other parts of the incineration clean-up equipment. This particulate matter, together with that from the filters and precipitators, has to be treated before landfill disposal (see 5.4 Waste from waste). The scrubbing solutions also have to be treated before ultimate disposal.

5.2.1 Types of incinerator

The choice of incinerator depends on factors such as the physical form of the waste, e.g. the waste may consist of drummed chemicals or disused parts of transformers, or more simply a mixture of liquids of different viscosities. A vital factor is the combustion characteristics of the waste, which include calorific value. The quantity of ash produced is an important consideration. Waste containing known hazardous constituents has to be specially appraised.

There are three basic types of incinerator: static chamber; moving hearth; and rotary kiln.

The static chamber incinerator is used for liquids or solids, which are fed directly on to the hearth or on to trays which are transferred into the chamber. Chemicals, contaminated soil, clinical waste and packaging materials are examples of materials which can be incinerated in this way. The main disadvantage of this method is the difficult removal of slag and bottom ash, despite constant raking of the burning waste.

The moving hearth incinerator is the type usually used for the incineration of municipal waste. It is similar to the static chamber incinerator, but as the name implies, the hearth moves so that the waste, as it burns, is transferred, with continuous agitation, from the front to the back of the chamber, the ash falling from the end of the conveyor.

The rotary kiln is the most efficient incinerator in that the waste is continually broken up by the rotating drum and the rotational movement enhances the turbulence and hence the mixing of air with the burning waste. Cement kilns can be used sometimes for this purpose.

An afterburner is frequently used to ensure the complete destruction of toxic combustible gases. This type of incinerator is the best for ensuring the destruction of infectious waste and for certain chemicals, e.g. PCBs, where a long residence time is necessary. The rotary kiln is the most expensive of the practical

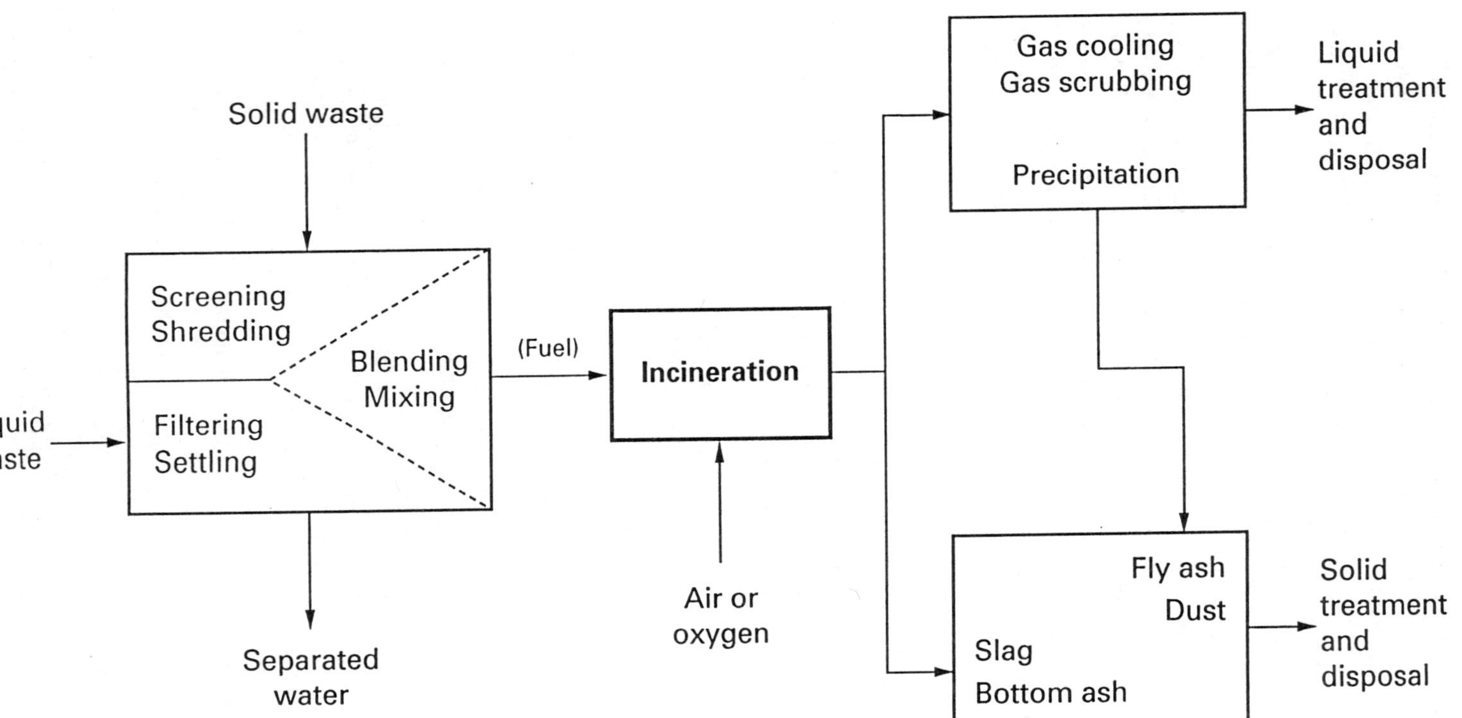

Figure 5   *Flow diagram of the incineration process*

incinerators, largely because some parts which wear have to be replaced at intervals, e.g. refractory linings. Most commercial incinerators, dedicated to the destruction of hazardous waste, are of the rotary kiln type, suitably adapted for specific needs. Such chemicals as PCBs, PCTs, chlorinated pesticides and halogenated solvents are incinerated in this way.

5.2.2 Emission control

In order to meet legal requirements and in order to increase the efficiency of incineration, emission control of gaseous pollutants is vital. Carbon dioxide and steam will always be gaseous products of the combustion of organic waste, but it is the toxic gases, viz. hydrogen chloride, fluorides, sulfur dioxide, oxides of nitrogen, and not least the products of incomplete combustion, e.g. chlorinated dioxins, chlorinated furans and polycyclic aromatic hydrocarbons, which have to be minimised.

Table 11 *Limiting values of emission in mg m^{-3} according to the nominal capacity of the incineration plant (89/369/EEC)*

Pollution	Less than 1 tonne per hour	From 1 tonne per hour to less than 3 tonnes per hour	3 tonnes and more per hour
Total dust	200	100	30
Heavy metals:			
Pb + Cr + Cu + Mn	–	5	5
Ni + As	–	1	1
Cd + Hg	–	0.2	0.2
Hydrochloric acid	250	100	50
Hydrofluoric acid	–	4	2
Sulfur dioxide	–	300	300

In 1989, the European Commission produced two Directives concerned with setting limits of concentration for gaseous pollutants emitted from the incineration of municipal waste. The first Directive, 89/369/EEC, covers new incineration plant and requires that the operation is carried out using the "best available technology not entailing excessive cost", (BATNEEC). In any event, certain emission limits have to be met and these are summarised in TABLE 11. The second Directive, 89/429/EEC, applies to existing incineration plant and sets time targets for the attainment of similar emission limits as for the new plant Directive. No emission limits have been set for chlorinated dioxins and chlorinated furans, each Member State having to fix their own, if so desired.

There is a proposed EC Directive concerned with setting limits for gaseous emissions from hazardous waste incineration (92/C130/01). This Directive is expected to have stringent limits for emissions of materials catalogued in the 1989 Directives, e.g. total dust, hydrochloric acid, hydrofluoric acid, sulfur dioxide and heavy metals such as lead, chromium, copper, manganese, nickel, arsenic, cadmium and mercury. The hazardous waste incineration Directive is also expected to include details of the ways of determining the emission limits and the analytical techniques necessary for the measurement of the pollutants in air.

There is considerable political, public and scientific interest in the levels of the products of incomplete combustion, e.g. chlorinated dioxins, some of which may be extremely toxic.

Chlorinated dioxins are formed from the incomplete combustion of certain chlorinated organic compounds, particularly PCBs and PCTs. An incineration temperature of at least 1200 °C is necessary to destroy PCBs and PCTs, and if the gaseous effluent stream is not cooled quickly to below 300 °C there is the possibility of the formation of chlorinated dioxins. If emissions of chlorinated dioxins or chlorinated furans occur, the environment generally, including the land surrounding the incinerator, may become contaminated.

The acidic emissions such as hydrogen chloride, volatile fluorides and sulfur dioxide are scrubbed with alkaline solution and the grit and dust are arrested by various mechanical and electrical devices. The fate of these residues is discussed in Chapter II 5.4 Waste from waste.

5.2.3 Ash disposal

Before contemplating disposal of incinerator ash, re-use or recycling should be considered. In some cases, ash can be processed to recover ferrous metal, and ash has been used as a low-grade aggregate in the construction industry.

In well-managed municipal waste landfill disposal sites, the waste should be covered over at the end of the working day to prevent attraction of vermin and birds and also to prevent loose paper etc from blowing away. Relatively metal-free incinerator ash can be used for such a purpose, which is economically sensible when an authority is running both an incinerator and a landfill facility. If the ash has a relatively high metal content, particularly lead and cadmium, the ash may be deposited in landfill sites which do not have a leachate problem. Otherwise, the ash can be physically and/or chemically treated before ultimate disposal in a landfill site. (see Chapter II 5.4 Waste from waste).

5.3 Landfill

Nearly all forms of waste disposal end with landfill, whether by direct disposal, disposal of treated waste; or residual waste from another process, e.g. incineration. A well-managed landfill operation will result in the controlled deposit of waste to land, (below or above surface level), in such a manner that the risk of polluting water, air and adjoining land is minimal.

Provided the sites are available and can be managed in a proper manner, landfill is often the first option to be considered for the disposal of municipal, commercial and non-hazardous industrial wastes. Treatment technologies applied as part of sound industrial practice will result in sludges, ash, filter cake, neutral inorganic solids, etc. for which no further treatment is practicable, other than landfill.

Because landfill is the cheapest method of waste disposal, approximately two thirds of European waste is deposited in this way. However, with the forthcoming impact of the proposed EC Directive concerning landfill, which will introduce stricter controls, the cost of landfilling is expected to rise considerably.

5.3.1 Landfill practice

The design and construction of landfill sites should be carried out in such a manner as to produce a facility which is efficient in practice and does not harm humans, animals or the environment; this means that there has to be control over the short and long term formation of the products of waste decomposition, i.e. leachate and landfill gas.

The proposed EC Directive on the landfill of waste (91/C 190/01) came as a result of concern about the poor standards of landfill generally prevailing in Europe. There was also a desire to harmonise conditions and criteria for landfilling. The proposed Directive defines landfill as "a site of waste elimination used for the controlled waste deposit on or in the earth" and it separates three types of landfill, viz.

- for hazardous waste
- for municipal and non-hazardous wastes, and for other compatible wastes as defined in the compatibility criteria set out in an Annex entitled, "Waste acceptance criteria and procedures"
- for inert waste

The acceptability criteria include details of control procedures and waste sampling for testing; and there is an elution test procedure included, which is a modification of DIN 38414-S4. The new Directive is expected to list assignment values for hazardous and inert wastes according to the results of the elution tests, which will use approved analytical procedures. TABLE 12 lists the parameters earmarked for the elution tests.

The proposed Directive also sets up different authorisation procedures for the creation and exploitation of landfill sites, and prohibits some wastes, e.g. liquids (with some exceptions), infectious waste from medical and veterinary sources, flammables and explosives; waste will not be allowed to be diluted to conform. Management of landfill sites will continue for up to ten years after their closure.

Landfill sites can also be categorised either according to whether they are designed to contain the leachate produced or to encourage leachate loss by evaporation and also, particularly, by attenuation, e.g. the clean-up action by the naturally occurring materials surrounding the landfill.

A contained site is one where the leachate is prevented from escaping, e.g. clay-lined or lined with high-density polyethylene. Where containment is practised, the leachate is collected, e.g. by bunding around the site and pumping out, before either on-site or off-site treatment.

Table 12 *Parameters for elution tests in the proposed EC Directive on landfill of waste*

pH	Fluoride
TOC	Ammonium
Arsenic	Chloride
Lead	Cyanide
Cadmium	Sulfate
Chromium(VI)	Nitrite
Copper	Adsorbed organically-bound halogens
Nickel	Chlorinated solvents
Mercury	Chlorinated pesticides
Zinc	Lipophilic substances
Phenols	

Where "disperse and attenuate" is practised, the leachate is allowed to escape into the strata below the landfill. This has been a popular method, using the attenuating process, which can have a physical action, e.g. adsorption, filtration and dispersion, or a chemical action, e.g. acid-base neutralisation, oxidation/ reduction and precipitation. There will also be biodegradation occurring in the underlying strata, which will help in the attenuation process. Attenuating may result in either the effective removal of a pollutant or in a delay to its release into the environment. Even with a good knowledge of the hydrogeology of the site, there is always the risk of the attenuation not being effective enough to clean up the leachate.

Generally, there are three arrangements for depositing waste in a landfill – monodisposal, multidisposal and codisposal.

Monodisposal is where one homogeneous waste is deposited, e.g. a slurry or liquid from a constant source.

Multidisposal, the most common type for deposition of non- hazardous waste, is where different types of compatible waste are deposited, e.g. food and bricks.

Codisposal is the mixing together of at least two types of different waste in a landfill, with the objective of using the activity of one component to degrade the others, e.g. there is a practice whereby municipal waste is allowed to mature for some months before placing a small quantity of hazardous waste within it, such that biodegradation or decomposition of the hazardous waste is encouraged.

5.3.2 Biodegradation

Municipal waste, especially that which has been freshly deposited, contains a large proportion of putrescible material, e.g. food and paper. Such waste undergoes biodegradation, i.e. the action of microbes to break down the waste into more simple organic compounds, such as carboxylic acids, and ultimately to water, carbon dioxide, hydrogen, methane and ammonia, depending on whether or not the microbial action is aerobic or anaerobic. Aerobic biodegradation takes place initially in landfilling and lasts until all the entrapped oxygen has been consumed when anaerobic biodegradation takes over.

The liquor resulting from biodegradation makes up the main proportion of landfill leachate (see also Chapter II 5.3.3 Groundwater pollution). The liquor has a high chemical and biochemical oxygen demand and often has polluting concentrations of ammonia and sulfide. Heavy metals, chlorides and mineral oils will invariably be present in the liquor. Even after dilution of the liquor by rainwater and surface water, the result is a strongly polluting liquid, which must be prevented from reaching the groundwater.

Mature landfilled municipal waste is not as biodegradable as fresh waste but the liquor from it can contain concentrations of ammonia and iron which need chemical treatment before ultimate disposal (see Chapter II 5.4 Waste from waste).

Biodegradation is encouraged in the "activated sludge" process (used extensively in sewage treatment) where micro-organisms in the sludge biodegrade the liquid waste.

5.3.3 Groundwater pollution

Most landfills, excepting those holding only inert waste, form leachates from a combination of rainwater, surface water and the organically strong liquor produced by the biodegradation of putrescible waste. The formation and composition of the leachate depends on the type and age of the waste, the dilution factor of the water and the extent of the physical and chemical alterations to the waste, as well as the extent of the biodegradation.

Where hazardous waste has been deliberately landfilled in a codisposal site, the leachate would have to be contained for treatment. Nevertheless, leachate from municipal waste landfill, in addition to the biodegradable products (see Chapter II 5.3.2 Biodegradation), can contain such substances as detergents, pesticides and chlorinated hydrocarbons, which, even if in low concentration, could have a polluting effect on groundwater if allowed to reach it.

The authorities who abstract water for drinking purposes exercise control over the quality of groundwater and have the duty to ensure that it does not get polluted by landfill leachate. This is a requirement of the EC Directive on the protection of groundwater against pollution caused by certain dangerous substances (80/68/EEC), which sets out conditions to protect aquifers from various sources of pollution, including landfill leachates. Annexed to the Directive is a list of substances which must be prevented from entering groundwater (List 1), and a list of substances which must be limited so as not to have a polluting effect on groundwater (List 2). These substances are shown in TABLE 13.

Pollutants in leachate should be qualitatively or quantitatively determined, as appropriate, on a regular basis. This will check whether or not there has been fly-tipping of any unscheduled waste which has the ability to pollute the leachate further.

The substances sought in the chemical analysis of the leachate are those listed in the EC Directive. The all-embracing, less specific tests of pH, suspended solids, conductivity, BOD, COD and TOC are very useful to perform before moving to the specific analyses.

Table 13 *Protection of groundwater against pollution EC directive 80/68/EEC*

List 1   Substances to be prevented from entering groundwaters

---

Organohalogen compounds and substances which may form such compounds in the
   aquatic environment

Organophosphorus compounds

Organotin compounds

Substances which possess carcinogenic, mutagenic or teratogenic properties in or via the
   aquatic environment

Mercury and its compounds

Cadmium and its compounds

Mineral oils and hydrocarbons

Cyanides

List 2   Substances to be limited so as to avoid pollution

The following metalloids and metals and their compounds

Zinc	Tin
Copper	Barium
Nickel	Beryllium
Chromium	Boron
Lead	Uranium
Selenium	Vanadium
Arsenic	Cobalt
Antimony	Thallium
Molybdenum	Tellurium
Titanium	Silver

Biocides and their derivatives not appearing in List 1

Substances which have a deleterious effect on the taste and/or odour of groundwater, and
   compounds liable to cause the formation of such substances in such water and to render
   it unfit for human consumption

Toxic or persistent organic compounds of silicon, and substances which may cause the
   formation of such compounds in water, excluding those which are biologically harmless
   or are rapidly converted in water to harmless substances

Inorganic compounds of phosphorus and elemental phosphorus

Fluorides

Ammonia and nitrites

There must be constant monitoring of the groundwater, using the same
analytical list as for leachate. If there has been pollution from the leachate, there
will be a necessity for the analytical methods to be more sensitive. The
groundwater will have been analysed in the pre-commissioning of the site and will
continue for up to ten years after landfilling operations have ceased.

5.3.4 Gaseous emissions

Gaseous emissions result as waste is decomposed and biodegraded. The decomposition process is complex but, in general terms, carbon dioxide and hydrogen are generated early in the decomposition. As the available oxygen is used up and the waste matures, the anaerobic degradation produces methane, while the carbon dioxide concentration falls. The more moisture there is in the waste, the greater its decomposition and the greater is the gas generation. The general consensus is that landfill gas from a mature waste is composed of 50-60% methane and 35-40% carbon dioxide, with small amounts of other gases, noticeably the odorous sulfides and mercaptans.

Although the methane may be collected and used as a fuel, its presence within the voids and channels of a landfill presents a risk to humans and the environment. Methane forms an explosive mixture with air, and venting and flaring are techniques used to eliminate it, and in so doing, remove the risk of explosion and fire. The construction of a new landfill site has to take into account the underlying strata and the formation of landfill gas and allow for its safe collection or destruction.

Methane and carbon dioxide may be monitored using portable detection equipment or the more accurate and specific on-site gas chromatographic determination. Samples of landfill gas may also be taken for laboratory analysis.

5.4 Waste from waste

It is impossible to eliminate waste completely. Incineration reduces the volume of waste, but there are products of combustion, e.g. ash and gases, which are generated. Landfill is waste's graveyard, i.e. waste from other processes, such as incineration and physical treatment, has its final resting place there. However, landfill produces its own waste in the form of leachate and evolved gas, both of which have to be dealt with to remove or minimise their environmental impact.

5.4.1 Waste from incineration (see also Chapter II 5.2 Incineration)

Waste resulting from incineration includes bottom ash plus slag and particulate from the cleaning of the gaseous emissions.

Bottom ash can be washed with water to remove water-soluble components, such as alkali metal salts, and the remaining ash can be disposed of at a landfill site, either by codisposal with municipal waste or as a cover material. Slags are usually impervious to water and can be disposed of directly to landfill.

Fly ash and the other particulates, arrested in cyclones and precipitators, can contain toxic substances. Care has to be taken with the handling of the particulates because of their fineness.

Sulfur dioxide, hydrogen chloride and volatile fluorides from incineration of waste are trapped in scrubbers containing alkaline solutions, which are based on sodium hydroxide, sodium carbonate or lime. The solutions are, necessarily, kept alkaline during the operation and the final solutions will contain sodium or calcium salts with excess alkali. These solutions are neutralised, diluted and desalinated as necessary before discharge to a watercourse or sewer.

5.4.2 *Waste from landfill (see also Chapter II 5.3 Landfill)*

Most landfill sites produce leachate which needs attention. That leachate which contains relatively low concentrations of chemicals and other substances may be allowed to percolate through the underlying strata, which will provide attenuation. Leachate, which contains significant concentrations of either inorganic species, such as metal salts, or organic substances of a high biochemical oxygen demand, must be treated before it is deposited. It is very unlikely that such a liquid can be discharged to a watercourse.

Leachate recirculation is practised in some countries and this entails circulation of leachate back through the waste which generated it. As an alternative, the leachate may be sprayed on to the land surrounding the site. The microbial action of the waste or the soil, respectively, can aid the cleaning of the leachate, albeit in a limited way.

A relatively high concentration of ammonia occurs in leachate from the degradation of putrescible waste in landfill and this can be reduced considerably by aeration with agitation of the leachate in the pH range 10.5–11.5.

The final resort is treatment of the leachate either on-site or after it has been transported to its ultimate destination, usually a sewage works. Treatment can be either chemical, e.g. precipitation, oxidation/reduction; or physical, e.g. adsorption; or biological, e.g. aerobic or anaerobic biodegradation. The aim is to produce a liquid clean enough to be accepted for discharge to a watercourse or sewer.

Landfill gases are waste products of landfill. In low concentration they are allowed to escape into the environment, but at higher concentration, methane, for example can be collected and used as a source of energy.

Chapter III

Legal, Administrative and Economic Aspects

1 Legal and administrative aspects

In any country or state, the Government, being the ultimate responsible authority, should provide an effective legal and administrative framework for waste control. There are variations in national approaches to the organisation and planning of waste management, but where the management is carried out in a proper manner there are certain essential common elements which are covered in this Chapter. It must be always born in mind that, in the longer term, environmental standards are likely to be made more stringent and allowance should be made for this possibility when designing waste handling facilities.

1.1 Identification and classification of waste

Under the EC Directive 91/156/EEC a list containing 645 named wastes has been catalogued. These wastes are classified into 20 categories which are based on the industries or activities of their origins. The EC Directive 91/689/EEC lists types and properties of hazardous ingredients of wastes together with hazardous properties of wastes.

Waste can also be classified in a more general way. e.g. municipal, commercial, industrial, clinical. Municipal waste and the commercial waste from offices, shops, restaurants, etc. would not normally be expected to contain hazardous substances or be of hazardous character. On the other hand, while the greater proportion of industrial waste is innocuous, there are some wastes which are known to be hazardous and others which need investigation for classification purposes.

1.2 Hazardous waste

Hazardous waste usually comes from industrial sources. The producer of the industrial waste should know what substances it contains and can pass the information on to the handler and disposer. This knowledge is gained, usually, by a study of the whole manufacturing process. Questions demanding an answer include:

*   what is the composition of the raw materials, how pure are they and what and how much are the impurities?
*   what is the final product and what are the by-products?

- how much and what type of waste is produced and at what point in the production?
- are there any hazardous constituents expected in the waste?

The next useful step may be appropriate, reliable chemical analysis. Full analysis of hazardous waste can be difficult and often impossible. It may not be necessary to carry out a full analysis provided it is possible to characterise the waste according to one or more of the properties listed in the hazardous waste Directive.

Proper sampling of waste streams for analysis is a prerequisite for reliable interpretations of analytical data. Statistically representative samples of sufficient number and size should be taken.

For chemical analysis, the waste must be in a liquid state. This means that solid material has to be extracted (or ashed and the ash extracted) using water, acid or organic solvent. FIGURE 6 is a flow diagram showing this. The resulting solution can be analysed for specific analytes where these are thought to be present, otherwise more general testing is required, at least to characterise the ingredients.

The combination of testing and analysis should yield sufficient information to judge whether or not a waste should be classified as hazardous and also for deciding on the appropriate method of treatment before disposal.

1.3 Waste management plans

In order to provide a framework for the attainment of sound waste management, plans should be drawn up in the first instance. The EC Directive 91/156/EEC on waste requires that Member States produce such plans, which should relate to:

- the type, quantity and origin of waste to be recovered or disposed of
- general technical requirements
- any special arrangements for particular wastes
- suitable disposal sites or installations

The plans are expected to cover such details as:

- the natural or legal persons empowered to carry out the management of waste
- the estimated costs of the recovery and disposal operations
- appropriate measures to encourage rationalisation of the collection, sorting and treatment of waste

Member States are expected to co-operate with each other, as necessary, on the formulation of the plans. They can also take measures to prevent waste movement should this conflict with the plans.

The hazardous waste EC Directive 91/689/EEC provides for hazardous waste to be included in waste management plans. Collaboration between Member States is particularly encouraged on planned methods of recovery and disposal of such waste.

Although recycling and waste treatment form important parts of waste

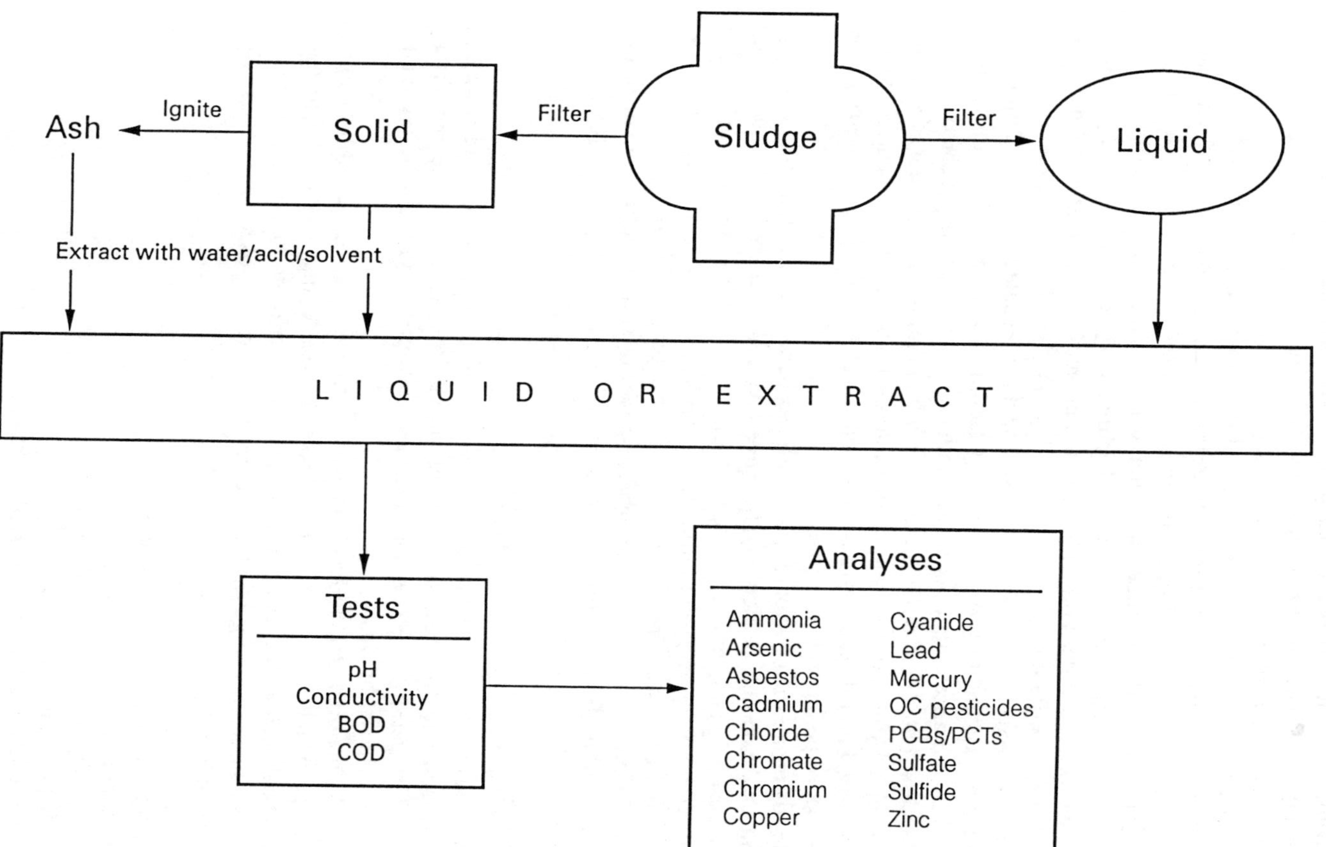

Figure 6　*Transformation of wastes into liquids for testing and analysis*

management, plans are often referred to as waste disposal plans because of the emphasis on this part of waste management. The waste disposal plans are managed at local level by regulatory authorities.

Waste disposal plans have to take account of the total amount of waste arising and set this against the total capacity, in terms of landfill space and/or incinerator time, to deal with it in an environmentally sound manner. Plans should include an accurate forecast of the volume of future wastes arising and also of capacity. Forecasting future requirements entails the regulatory authority working closely with the relevant industries and with the planning authority, particularly when new facilities have to be found.

Waste disposal plans should also include strategies for improving environmental acceptability of the existing facilities. For example, with existing landfill sites, nuisances, such as traffic pollution and noise, may be reduced by altering access routes, and any malodours may be eliminated by speedier covering of putrescibles. Aftercare and site restoration are essential requirements in waste disposal plans. Landfill sites have to be cared for after tipping has ceased at least until the risk of pollution is minimal.

Waste disposal plans should take into account the costings of recycling, treatment and disposal, particularly the investment needed for incineration plant and the cost of land earmarked for landfill sites (there can be considerable competition for land-use). Pollution abatement technology is another economic consideration (see under Chapter III 2 Economic aspects).

Waste is expected to be managed and disposed of within the region where it is produced and this should be reflected in the plans.

1.4 Waste permits and records

The regulatory authorities which draw up waste disposal plans are those which are usually responsible for the issue, control and enforcement of waste permits (or licences). The permits are granted to those who fulfil certain conditions for the recovery or disposal of waste.

The EC Directive 91/156/EEC on waste requires that any establishment or undertaking, which carries out any of the 15 waste disposal or 13 recovery operations annexed to the Directive, must obtain a permit to do so. A permit is not required if the waste is generated and recovered or disposed of in-house. The permits have to cover:

- the types and quantities of waste
- the technical requirements
- the security precautions to be taken
- the disposal site;
- the treatment method, where applicable.

Permits are granted for specified periods and may be renewable; they may be subject to conditions and obligations and may be refused should the operation be unacceptable or not environmentally friendly. The Directive also requires professional waste recoverers and disposers to be registered with the authority. Similarly, the EC Directive 91/689/EEC on hazardous waste requires registration for operators.

The regulatory authority, prior to the issue of a permit, must be satisfied that the applicant is preparing to operate, or is already operating in an environmentally acceptable manner. The permit, once granted, may be suspended or revoked if the owner subsequently contravenes or by-passes any of the approved conditions under which the permit was first granted.

A landfill permit includes conditions relating to waste input, control of leachate and landfill gas, and control of noise, dust and litter. Conditions of permits for other facilities also take into account that waste will probably leave the site, either recovered or treated.

The conditions which should be included in waste permits are shown in TABLE 14. The site must be properly prepared, particularly to give environmental control; the infrastructure, including security, must also be in place before operations commence. The types and quantities of permitted wastes are essential information. The working plan of the site, included with these conditions, must be maintained, regularly up-dated and available for inspection at all times.

Table 14 *Some conditions to be included in a waste permit*

Site preparation
Site infrastructure
Types of waste
Quantities of waste
Operation of site
Plant maintenance
Pollution monitoring and control
Post-operational control
Records
Site inspections
Emergency procedure

A system to tackle emergencies must be included in the permit and there should be a plant maintenance schedule. A breakdown in large machinery, such as that used in incineration, can cause environmental problems as well as financial loss, and the risk of this happening should be minimised. In any event, pollution monitoring and control measures must be present in the permit.

One of the most important conditions of the permit is the obligation to record all information relating to the site. EC Directive 91/156/EEC on waste requires establishments or undertakings to keep a record including the quantity, nature and origin of the waste, and, where relevant, the destination, frequency, collection, mode of transport and treatment of waste.

Site inspections by the regulating authority should be made at frequent intervals. The outcome of inspections are recorded and any irregularities in the conditions must either be rectified or the permit is withdrawn.

2 Economic aspects

Although the key to good waste management is the efficient conduct of the necessary operations, with no harm to human health or pollution of the environment, these operations should be completed in the most economic manner.

Waste reduction and recycling can result in considerable financial saving. Apart from cutting production costs, the reduction in long-term environmental liability and handling hazards result in lower insurance premiums.

Recycling is not only environmentally friendly but can also be profitable. Profit is possible if the new products can be produced economically, and provided there is a consistent market for them. Many recycling schemes, particularly those involving the public's help, for example, in separating and saving items of glass, paper, plastic and metalware can be profitable, although there are occasions when the market for recycled materials may be depressed (see under Chapter II 1.3.2 Marketability and economics of recycling).

Large savings can be made by not having to treat or dispose of waste, e.g. by avoiding the labour and administration costs of collecting and delivery, costs of treatment and disposal, capital outlay on purchasing the site and equipment, operational costs, and, for landfill, the expense involved in after-care.

The costs of a recycling scheme will include some or all of the following:

- labour and administration costs of collecting waste and delivering recyclables to the processing facility, which will transform the recyclables into marketable products
- labour and administration costs of the processing
- capital costs of the recycling equipment
- operating costs of the processing, including power, fuel, maintenance and insurance
- administration costs of promoting the scheme and marketing the products

The financial appraisal should take account of a life cycle analysis, which is a management tool for the purpose (see under Chapter III 3.2 Life cycle analysis).

Economics must play an important part in the decision-making process concerning the choice of waste disposal methods. This is evident, particularly, in municipal waste disposal where the relative costs of landfill and incineration will vary at different sites and must therefore be evaluated.

In regions where there are available "holes in the ground", e.g. exhausted quarries and natural field depressions, landfilling can be accomplished economically, since landowners may wish to return their land to agricultural use and be paid for so doing. Where land prices are high or where land is not available for landfill, incineration may be the only alternative option, given that exporting the waste is effectively discouraged by the Basel Convention (see under Chapter II 3.1). The capital cost and maintenance of incineration plant is very high compared to similar cost in relation to landfill operation and it is generally accepted that incineration is the more costly method of disposal. However, with land prices expected to go on rising, together with fiscal policies to discourage landfilling, the differential between the two methods is expected to narrow, although the commercial benefit of the methane, generated in landfill, has yet to be fully realised.

Where permitted, liquid industrial wastes and also landfill leachates can be disposed of to sewer. This option can be more cost-effective than on-site treatment and then disposal.

Greater awareness by both industry and public of the benefits of waste reduction and recycling has had a positive effect on the economic aspects of waste management.

3 Management tools

3.1 Waste audits

Responsible industries use environmental impact assessment as an instrument of their integrated pollution control (IPC) programme. Most industries generate waste, the environmental effects of which have to be considered in the environmental impact assessment. It is necessary, therefore, to have a detailed knowledge of the wastes arising and the way in which they are handled, stored and treated; this is achieved by waste audit. Small and medium enterprises might not make an environmental impact assessment but, if they are waste generators, they should have a waste audit system.

The waste audit is not simply a list of types and quantities of waste. It is a system of checking that the documentation and records match what is actually taking place.

When waste is generated it will be stored awaiting collection for recycling, treatment and/or disposal, and the safe handling of the waste should be audited from this point. The audit includes checking that the waste is in the correct containers and that it is segregated where necessary, e.g. hazardous from non-hazardous waste. The labelling on the containers should be checked for accuracy and durability; also the storage time should be checked in case this has been exceeded.

The records should be checked to see if the detail is adequate and whether or not policy objectives have been met, e.g. waste reduction and recycling targets. Certain administrative procedures and costings may also form part of an audit. The emphasis is on safety, and the emergency procedures should be checked to see how efficiently any past emergency has been controlled and what steps have been taken to improve on that record. Auditing should also cover safety training records of the workforce.

All waste audits should be properly documented and any shortcomings should be rectified as promptly as possible. The quality of the audit itself may be assessed by a review body.

There is apparent overlap between the record keeping (see under Chapter III 1.4 Waste permits and records) and waste auditing, but the distinction is that auditing should be carried out by qualified persons who are independent of the staff whose functioning is under scrutiny.

3.2 Life cycle analysis

Life cycle analysis (LCA), sometimes known as cradle-to grave analysis, is a management tool which integrates the various stages of operations with environmental impact, from raw materials through production to disposal.

In waste management, LCA is used, primarily, to reveal the environmental impacts of an operation which may not otherwise have been taken into account. A secondary consideration is whether or not the re-use or recycling proves to be a more economically viable option than using the waste as an energy source or simply disposing of it.

Given adequate education, there will be environmental motivation behind recycling of glass, paper, plastics and aluminium shown by the public's enthusiastic participation in taking waste to the appropriate banks. LCA may show that such recycling is not environmentally viable. The example of purposely driving a car to deliver empty bottles to a bottle bank has been referred to in Chapter II 1.3 Recycling.

In LCA, every aspect is looked at which adds to or subtracts from the overall environmental and economic state. Paper recycling is an example of how useful LCA can be. The paper "waste" is transferred to the mill for re-pulping, which requires fuel for the collection and transportation. The re-pulping process requires energy and there is a resulting residue which has to be disposed of. Any inks or dyes present have to be removed chemically at this stage. The pulp is made into new articles and these have to be packaged and transported to the customer; all these operations require fuel or energy input.

The revenue from the marketed articles plus the additional savings in primary raw materials through not having to treat and dispose of the waste is set against the total expenses. The calculation confirms or otherwise the profitability of the recycling. There is also the alternative that a LCA may show that burning the paper as a fuel is the most economic option. Notwithstanding the economics, the operation must not show an environmental dis-benefit.

One of the most significant costs in re-use and recycling is transportation. A proper appraisal of transport costs must include the depreciation of the vehicles and road costs, for example. Re-use or recycling carried out at the same site as that for production, with the resulting saving in transport costs, can make the difference between a profitable re-processed product and a loss-maker.

The re-use and recycling of plastic containers is widely practised and the technology of the operations involved has improved with time, in part as a result of LCA. Where costs of some operations have been shown to be too high, one successful cost-reducing ploy has been to reduce the thickness of the plastic, thus reducing the weight per item. Not only has this resulted in a saving in the cost of production but also in transportation, for example.

Cars and some domestic appliances can be made using components which are known to be economically recyclable, once the useful life of the item is over. This means that the manufacturer will happily take it back from the customer and everyone gains – the manufacturer makes a profit and the customer does not have to dispose of the item and thus add to the environmental burden.

Re-use and recycling are ideals which are usually environmentally commendable. The strength of LCA is in showing where all the environmental and economic costs are, so that systems may be changed to reduce some of these costs.

References

European union publications

'A Community Strategy for Waste Management' – A Communication from the Commission to the Council and to Parliament. 18 Sept. 1989 (This document is also found within the following reference).

European Community Environment Legislation; Vol.6 "Waste" CEC-DG XI: Environment, Nuclear Safety and Civil Protection, 1992

'Commission Decision 94/3/EEC of 20 Dec. 1993 establishing a list of wastes pursuant to Article 1(a) of Council Directive 75/442/EEC on waste'. *Official Journal of the European Communities*, L5/15, 7 January 1994

Council Directive of 17 December 1979 on the protection of groundwater against pollution caused by certain dangerous substances (80/68/EEC). *Official Journal of the European Communities*, L20/43, 26 January 1980

Council Directive of 18 March 1991 amending Directive 75/442/EEC on waste (91/156/EEC). *Official Journal of the European Communities*, L78/32, 26 March 1991

Council Directive of 12 December 1991 on hazardous waste (91/689/EEC). *Official Journal of the European Communities* No. L377/20, 31 December 1991

Council Regulation (EEC) on the supervision and control of shipments of waste within, into and out of the European Community. Council Regulation No.259/93, 1 February 1993. *Official Journal of the European Communities*, L30/1, 6 February 1993

Council Resolution of 7 May 1990 on Waste Policy, 90C, 122/02. *Official Journal of the European Communities*, C122/2, 18 May 1990

'Proposal for a Council Directive on the Landfill of Waste'. Issue C190, *Official Journal of the European Communities*, C190, 22 July 1991

UK government publications

Department of the Environment. Waste Management Paper No. 1. 'A Review of Options. A memorandum providing guidance on the options available for waste management treatment and disposal'. Department of the Environment, Wastes Technical Division, London, 1992

Department of the Environment. Waste Management Paper no. 2/3. 'The Preparation of Waste Disposal (Management) Plans'. Draft for Consultation. Department of the Environment, London, 1993 September

Department of the Environment. Waste Management Paper No. 4. 'The Licensing of Waste Management Facilities'. HMSO, London, 1994

Department of the Environment. Waste Management Paper No. 6. 'Polychlorinated Biphenyl (PCB) Wastes'. HMSO, London, 1976

Department of the Environment. Waste Management Paper No. 7. 'Mineral Oil Wastes'. HMSO, London, 1976

Department of the Environment. Waste Management Paper No. 26. 'Landfilling Wastes. [A technical memorandum for the disposal of wastes on landfill sites]'. HMSO, London, 1986

Department of the Environment. Waste Management Paper No. 28. 'Recycling. A memorandum providing guidance to local authorities on recycling'. HMSO, London, 1991

Department of the Environment. 'Waste management: The Duty of Care, A Code of Practice'. HMSO, London, 1991

House of Lords Select Committee on the European Communities. 'Remedying Environmental Damage (with minutes of evidence)'. HMSO, London, 1993

Other publications

ACS Task Force on RCRA. 'Less is Better. Laboratory Chemical Management for Waste Reduction'. American Chemical Society, 1985 [RCRA = Resource, Conservation and Recovery Act (USA)]

British Medical Association. 'Hazardous Waste and Human Health'. Oxford University Press, 1991

L.W. Canter and R.C. Knox. 'Ground Water Pollution Control'. Lewis Publishers Inc., Chelsea, Michigan, USA, 1985

CEFIC. Guidelines for the Protection of the Environment. 'Industrial Waste Management. A CEFIC Approach to the Issue'. CEFIC, 1987.

M.A. Curran 'Broad-based Environmental Life Cycle Assessment'. *Environ. Sci. Technol.*, 1993, **27**(3), 430

C.R. Dempsey and E.T. Oppelt. 'Incineration of Hazardous Wastes: A Critical Review Update'. *J. Air Waste Manage. Assoc.*, 1993, **43**, 25

ECETOC. Technical Report No. 49. 'Exposure of Man to Dioxins: A Perspective on Industrial Waste Incineration'. ECETOC, Brussels, 1992

Environmental Resources Ltd (for WHO Regional Office for Europe). 'Identification of Priority Chemicals in Hazardous Wastes. World Health Organisation Regional Office for Europe, 1990

European Research and Consulting sprl (ERCO) (for the European Foundation for the Improvement of Living and Working Conditions). 'Transport of Dangerous Wastes'. European Foundation for the Improvement of Living and Working Conditions, Shankhill, Ireland, 1987 (Obtainable from the Office for Official Publications of the European Communities, Luxembourg)

J.J. Ferrada 'Hazardous Chemical Waste Management'. Garland Publishing, New York and London, 1990

K. Fouhy *Chem. Eng. (N.Y.)*, 1993, **100**(7), 30. 'Life Cycle Analysis Sets New Priorities'

R.M. Harrison (ed.). 'Pollution: Causes Effects and Control'. The Royal Society of Chemistry, Cambridge, 1990

R.E. Hester and R.M. Harrison (eds). 'Waste Incineration and the Environment'. Issues in Environmental Science and Technology, vol. 2. The Royal Society of Chemistry, Cambridge, 1994

T.E. Higgins 'Hazardous Waste Minimization Handbook'. Lewis Publishers, Chelsea, Michigan, 1989

ISWA. *Journal of the International Solid Wastes and Public Cleansing Association*, 1994 **12**(3) 'Waste Management & Research'. Academic Press

Lave, L.B. *et al. Environ. Sci. Technol.*, 1994 **28**(1), 19A. 'Recycling Decisions and Green Design'.

J. P. Lehman 'Hazardous Waste Disposal' Plenum Press, New York, 1983

S.E. Manahan 'Hazardous Waste Chemistry, Toxicology and Treatment'. Lewis Publishers, Chelsea, Michigan, 1990

K. Martin and T.W. Bastock (eds). 'Waste Minimisation: A Chemist's Approach'. Royal Society of Chemistry, Cambridge, 1994

R.B. Mitra 'Hazardous Waste Management in Developing Countries' Central Leather Research Institute, Adyar, Madras, 1989

J. Moller (ed.). 'International Directory of Solid Waste Management 1993/4: The ISWA Yearbook'. The International Solid Waste and Public Cleansing Association, 1993

R.W. Phifer and W.R. McTigue, Jr. 'Handbook of Hazardous Waste Management for Small Quantity Generators'. Lewis Publishers, Inc., Chelsea, Michigan, 1988

N.C. Singh *Ecodecision,* Jan. 1994, 'Integrating Economic Development and Waste Management – In Small Island States'

N.C. Singh 'Waste Management and Sanitation: Integrated and Co-ordinated Approaches to Solutions'. Paper presented to the Tenth Commonwealth Health Ministers Meeting in Cyprus, October 1992. CHMM92/TT/2.1 Commonwealth Secretariat, Marlborough House, London

J. Skitt (ed.). '1000 Terms in Solid Waste Management'. International Solid Wastes and Public Cleansing Association, Copenhagen, 1992

W.S. Sloan 'Site selection for New Hazardous Waste Management Facilities'. WHO Regional Publications, European Series No.46. WHO Regional Office for Europe, Copenhagen, 1993

M.J. Suess (ed.). 'Solid Waste Management – Selected Topics'. WHO Regional Office for Europe, Copenhagen, 1985

M.J. Suess and J.W. Huismans, (eds.) 'Management of Hazardous Waste'. WHO Regional Publications, European Series No.14. WHO Regional Office for Europe, Copenhagen, 1983

Technicians Today, 1990, **2**,(1) 'Hazardous Waste Management in the Chemical Industry' [various articles]. American Chemical Society

D.C. Thomas (ed. K.A. Ream). 'Recycling'. ACS Information Pamphlet. American Chemical Society, 1993

UNEP. 'Basel Convention on the Transboundary Movements of Hazardous Wastes and their Disposal. FINAL ACT'. UNEP, 1989

UNEP Environmental Law and Institutions Unit. UNEP Environmental Law Library No. 2. 'The Basel Convention on the Control of Transboundary Movements of Hazardous Wastes and their Disposal'. UNEP, 1990

WHO. 'Glossary on Solid Waste'. WHO Regional Office for Europe, Copenhagen, 1980

WHO Expert Committee. 'Environmental Pollution Control in Relation to Development'. WHO Technical Report series 718. World Health Organisation, Geneva, 1985

World Wildlife Fund & The Conservation Foundation. 'Getting at the Source: Strategies for Reducing Municipal Solid Waste'. World Wildlife Fund & The Conservation Foundation, 1991.